Op-Amp Circuits and Principles

Op-Amp Circuits and Principles

Howard M. Berlin

A Division of Prentice Hall Computer Publishing
11711 North College, Carmel, Indiana 46032 USA

© 1991 by Howard M. Berlin

FIRST EDITION
SECOND PRINTING—1992

All rights reserved. No part of this book shall be reproduced, stored in a retrieval system, or transmitted by any means, electronic, mechanical, photocopying, recording, or otherwise, without written permission from the publisher. No patent liability is assumed with respect to the use of the information contained herein. While every precaution has been taken in the preparation of this book, the publisher and author assume no responsibility for errors or omissions. Neither is any liability assumed for damages resulting from the use of the information contained herein.

International Standard Book Number: 0672-22767-3
Library of Congress Catalog Card Number: 91-60374

Acquisitions and Development Editor: *James Rounds*
Manuscript Editor: *Judy J. Brunetti*
Technical Reviewer: *Stephen J. Bigelow*
Cover Design: *George Harris*
Illustrator: *Tom Emrick*
Indexer: *Sherry Massey*
Production: *William Hurley, Bob LaRoche, Dennis Sheehan, and Vicki West*
Compositor: *Shepard Poorman Communications Corporation*

Printed in the United States of America

Trademark Acknowledgments

All terms suspected of being trademarks or service marks have been appropriately capitalized. SAMS cannot attest to the accuracy of this information. Use of a term in this book should not be regarded as affecting the validity of any trademark or service mark.

Contents

PREFACE ix

Chapter 1

WHAT IS AN OPERATIONAL AMPLIFIER? 1
    Introduction and Objectives   1
    Fundamentals of the Ideal Operational Amplifier   2
    Power Supplies for Operational Amplifiers   4
    Operational Amplifiers with Negative Feedback   5
    The Inverting Operational Amplifier   5
    The Noninverting Operational Amplifier   7
    The Voltage Follower   9
    Summing Amplifiers   9
    The Difference Amplifier   13
    The Instrumentation Amplifier   16
    JFET-Input Operational Amplifiers   19
    CMOS Operational Amplifiers   19
    Device Identification   20
    Package Styles   22
    Some Precautions   23

Chapter 2

PRACTICAL PERFORMANCE CONSIDERATIONS 25
    Introduction and Objectives   25
    The Data Sheet   26
    Input and Output Impedances   34
    Contributions to Output Offset Voltage   37
    Output Offset Null Adjustments   43
    Common-Mode Rejection Ratio   46
    The Effect of Frequency On Circuit Performance   49
    Frequency Compensation   53

Chapter 3

DIFFERENTIATORS AND INTEGRATORS 55
    Introduction and Objectives   55
    The Differentiator   56
    Integrators   62
    Differentiation and Integration of Common Waveforms   66

Chapter 4

SINGLE SUPPLY OPERATION 69
    Introduction and Objectives   69
    Single Supply Biasing   69
    Voltage Follower   70
    Noninverting Amplifier   73
    Inverting Amplifier   75
    Inverting Summing Amplifier   78
    Difference Amplifier   78

Chapter 5

THE NORTON AMPLIFIER 83
    Introduction and Objectives   83
    The Norton Amplifier   83
    Biasing the Norton Amplifier   84
    Biasing with Low Supply Voltages   85
    Norton Inverting Amplifier   86
    Norton Noninverting Amplifier   88
    Norton Voltage Follower   89
    Norton Summing Amplifiers   90
    Norton Difference Amplifier   91

Chapter 6

NONLINEAR SIGNAL PROCESSING CIRCUITS 95
    Introduction and Objectives   95
    Precision Rectifiers   96
    The Peak Detector   97
    Comparators   100
    Sample-and-Hold Amplifier   113
    Logarithmic and Antilogarithmic Amplifiers   114

Chapter 7

VOLTAGE AND CURRENT REGULATORS 119
    Introduction and Objectives   119
    Series Regulators   120
    The Shunt Regulator   127
    Current Regulators   129

Chapter 8

WAVEFORM GENERATION.. 131
    Introduction and Objectives   131
    The Basic Oscillator   131
    RC Oscillators   132
    Square-Wave Generators   137
    Triangle-Wave Generators   140
    Sawtooth Generator   144

Chapter 9

ACTIVE FILTERS.. 145
    Introduction and Objectives   145
    Filter Responses   146
    Active Filter Circuits   147

Chapter 10

EXPERIMENTS... 165
    Introduction   165
    Performing the Experiments   165
    Experiment 1—Voltage Follower, Noninverting, and Inverting Amplifiers   168
    Experiment 2—The Summing Amplifier/Averager   172
    Experiment 3—The Difference Amplifier and Common-Mode Rejection   174
    Experiment 4—Measurement of Operational Amplifier Parameters   178
    Experiment 5—The Differentiator and Integrator   183
    Experiment 6—Single Supply Biased Inverting AC Amplifier   187
    Experiment 7—The Norton Amplifier   190
    Experiment 8—Precision Half- and Full-Wave Rectifiers   195
    Experiment 9—The Peak Detector   198
    Experiment 10—Operational Amplifier Comparators   201
    Experiment 11—The Phase-Shift Oscillator   207
    Experiment 12—2nd-Order Low- and High-Pass Butterworth Active Filters   209
    Experiment 13—Active Bandpass and Notch Filters   217
    Experiment 14—The State-Variable Filter   227

GLOSSARY.. 235

Appendix A

DATA SHEETS... 239

Appendix B

STANDARD RESISTOR AND CAPACITOR VALUES 277
    Resistors   277
    Capacitors   278

Appendix C

REQUIRED PARTS AND EQUIPMENT FOR THE EXPERIMENTS 279

INDEX 281

Preface

Of the integrated circuits that are available, perhaps the most widely used (if not the most versatile) is the operational amplifier. In many cases they are cheap, with some costing less than 50 cents in single quantities. They are easy to use and require only low-level calculations for most designs. This is probably why the present designs of many analog circuits are much simpler than they were decades ago when some of us were playing with vacuum tubes and transistors.

This book is about the design and operation of basic operational amplifier circuits, coupled with a series of 14 experiments that illustrate and reinforce the design and operation of linear amplifiers, differentiators and integrators, comparators, precision rectifiers, peak detectors, oscillators, regulators, filters, single supply AC amplifiers, and the Norton amplifier. This book is not meant to be a sourcebook of all available operational amplifier circuits, nor is it a rigorous college textbook. However, this is a text/workbook that outlines in an easy-to-read style the designs and operation of the fundamental circuits that are the building blocks of the more sophisticated systems using many operational amplifiers. For this reason, this book should prove most useful to the beginning experimenter and hobbyist who wants to learn the basics by self-study, or it can easily serve as an addition to those college courses on linear integrated circuits, particularly those which have a laboratory section.

In the first nine chapters the fundamental circuits using bipolar, JFET-input, CMOS, and Norton-type operational amplifiers are discussed, with many numerical examples worked out. In all cases, standard component values are used. The chapters are divided by overall function: linear amplifiers, differentiators and integrators, single supply operation, Norton amplifiers, nonlinear circuits, voltage and current regulators, waveform generation, and active filters. Although specialized integrated circuits are available for several of these functions (most notably the comparator and voltage regulator), several chapters discuss how they can be built using operational amplifiers.

Aside from the final chapter, each chapter has an explanation of the instructional objectives to be gained in that chapter. As further aids for the reader, each major equation is numbered, and the designs of many of the basic circuits are summarized in Design Facts, which include both the schematic as well as the necessary design information. The final chapter contains 14 step-by-step experiments that illustrate and reinforce many of the concepts presented in earlier chapters.

In addition to the worked Example and Design Fact summaries, a glossary of over 50 terms relating to operational amplifiers is included to aid the reader, most notably the beginner. Furthermore, appendixes include (1) tabulations of standard resistor and capacitor values, (2) data sheets for the following operational amplifiers: CA3140, µA747, LF351, LM101/201/301, LM218/318, and LM2900/3900, and (3) a complete parts list needed for performing the optional experiments.

In the preparation of this book, I would like to acknowledge the many helpful suggestions offered by Stephen J. Bigelow, as well as the assistance of RCA, Signetics, and Texas Instruments who permitted the reproduction of the data sheets of some of their products.

HOWARD M. BERLIN

Chapter 1

What Is an Operational Amplifier?

INTRODUCTION AND OBJECTIVES

The term *operational amplifier*, or *op-amp*, was originally applied to high-performance DC differential amplifiers that used vacuum tubes. These amplifiers formed the basis of the analog computer which was capable of solving differential equations. As early as 1948, the vacuum tube op-amp was capable of performing mathematical operations such as addition, subtraction, arithmetic sign inversion, averaging, scaling (i.e., multiplication or division by a constant), integration, differentiation, comparisons, logarithms, and inverse logarithms. Because of its size and power requirements, the vacuum tube soon gave way to the transistor, and eventually to the integrated circuit as improvements in technology were introduced. Except for a marked reduction in size and cost, the function of today's op-amp has changed very little from the original version.

Based on the early vacuum tube op-amps and improvements in integrated circuit technology, the first series of commercially available op-amps were made in the mid 1960s. In 1963, Fairchild Semiconductor introduced the first commercially available op-amp, the μA702. It was followed two years later by the μ709. In 1965, National Semiconductor had introduced the LM101, while Fairchild unveiled the ever-popular μA741 in 1967.

Instead of using individual transistors and wiring them with resistors, capacitors, and diodes, use an integrated circuit op-amp (such as the 741). It is an example of a *monolithic integrated circuit* as its internal equivalent components are fabricated from a single piece of solid-state material into an integrated circuit.

At the completion of this chapter, you will be able to:

- List the characteristics of the ideal op-amp.
- Describe how the leads of both DIP and TO-type integrated circuit devices are numbered.
- Briefly describe how to decode important information about a given device from its part number.
- Wire a bipolar or split power supply.
- Identify (by name) the input, output, and power supply terminals of a typical op-amp from its schematic symbol.
- Explain the differences between open-loop gain, loop gain, and closed-loop gain.
- Draw and determine the output voltage of the following circuits:
    noninverting and inverting amplifiers
    voltage follower
    inverting and noninverting summing amplifiers (adders)
    difference amplifier

FUNDAMENTALS OF THE IDEAL OPERATIONAL AMPLIFIER

The basic purpose of an electronic amplifier is to increase the size of a signal. Besides voltage, the input signal parameter to be increased may also be current or power. A *linear amplifier* not only increases the signal's level but also produces an output signal that is a faithful reproduction of the input. A small signal applied to the input terminals of a perfect linear amplifier will produce a larger signal of the same waveform at its output terminals.

The op-amp is a device that lends itself to the construction of very good linear amplifiers, as well as many nonlinear circuits. To keep things simple, this chapter first discusses the *ideal* op-amp and how it is used to form voltage followers as well as noninverting, inverting, summing, and difference amplifiers.

As stated previously, the op-amp originally was a term used to describe a high-performance DC differential amplifier using vacuum tubes that formed the basis of the analog computer. As a result of improvements in technology, today's op-amp is a solid-state integrated circuit device that uses external feedback networks to control its response.

The ideal op-amp would be expected to have the following major characteristics:

1. Infinite open-loop voltage gain.
2. Infinite input impedance.
3. Zero output impedance.
4. Infinite bandwidth.
5. Zero offset voltage (i.e., zero input produces zero output).
6. Infinite common-mode rejection.

In practice, no op-amp can meet these six ideal characteristics. In the next few chapters we shall see that it is still possible to achieve high-performance circuits despite the fact that there is no such thing as an ideal op-amp. Very often we are able to assume that we are working with ideal op-amps, as we will be doing in this chapter. The results that are obtained will usually be more than adequate for the particular application.

As shown by the two common schematic symbols of Figure 1-1, the op-amp has two input terminals. One input is called the *noninverting input* and is indicated on the schematic symbol as a plus (+) sign. The other input terminal is called the *inverting input* and is indicated by a minus (−) sign. A single output is shown to the right. The op-amp is usually powered by a dual-voltage or split power supply. The positive supply voltage with respect to ground is often designated as $+V_{CC}$ (i.e., the collector supply voltage) while the negative supply voltage with respect to ground is $-V_{EE}$ (i.e., the emitter supply voltage), although it is occasionally written as $-V_{CC}$. As almost all op-amp circuits perform the same regardless of the supply voltage, the actual power supply connections are often omitted from schematics for clarity as they are implied to have been made and must be included in an actual circuit.

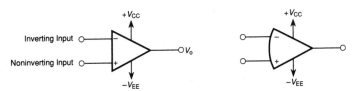

Fig. 1-1. Bipolar op-amp schematic symbols showing inputs, output, and power supply connections.

There is no standard convention as to which input is drawn above the other. For clarity some op-amp circuits will have the noninverting input on top. For others the inverting input is above the noninverting input. The left-hand triangular symbol is the preferred symbol, although the older symbol having a curved side is still used by some engineers as well as by some books and magazines.

3

The ideal op-amp basically functions as a difference, or differential amplifier. Consequently, it will amplify any difference in voltage that appears between these two inputs. Without any type of path that feeds back part or all of the output signal to one of its two inputs, the op-amp is said to be operating in the *open-loop mode*. The internal voltage gain of the op-amp in this case is termed the *open-loop gain* (A_{OL}). In the open-loop mode, the op-amp's output voltage equals the input voltage difference times the open-loop gain

$$V_O = A_{OL}(V_1 - V_2) \qquad \text{(Eq. 1-1)}$$

where,
V_1 = voltage at the noninverting input with respect to ground,
V_2 = voltage at the inverting input with respect to ground.

Equation 1-1 is a generalized equation where the actual polarities of V_1 and V_2 may be positive or negative with respect to the circuit ground. If both inputs are at the same voltage, this is referred to as the *common-mode input voltage* (i.e., the same voltage is *common* to both inputs), then $V_1 = V_2$ and the output is zero.

POWER SUPPLIES FOR OPERATIONAL AMPLIFIERS

Most op-amp circuits require a well-regulated dual, or split power source for proper operation. This is typically in the range of ±5 to ±15V. Depending on certain circumstances, these power supplies may be either driven by the AC power line or battery powered. However, in Chapter 4 we will see how to operate the op-amp for some applications with only a single supply voltage.

A split power supply (Fig. 1-2) has three terminals. One terminal is positive (+V_{CC}) with respect to the ground, while the other terminal is negative (−V_{EE}) with respect to ground. The third, or *common* terminal, must always be connected to the circuit ground. Depending on the complexity of a given circuit, these power supplies may be required to deliver from several milliamperes of current up to several amperes.

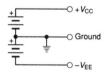

Fig. 1-2. Schematic of a dual, or split voltage power supply used with op-amps.

Often, the positive and negative power supply terminals are capacitively decoupled, or bypassed to ground. These capacitors (0.01 to 0.1 μF *ceramic* capacitors) are physically connected as close as possible to the op-amp supply pins. Capacitive bypassing provides an AC path to ground for the transfer of electrical noise between the power supply and the op-amp. Although it is not always absolutely necessary and is sometimes done partially or not at all, bypassing is excellent insurance against noisy and unstable op-amp operations.

OPERATIONAL AMPLIFIERS WITH NEGATIVE FEEDBACK

The gain that the op-amp provides without any external circuitry connected between its output and one of its inputs is called the open-loop gain (A_{OL}). For the popular general-purpose type-741 op-amp, the open-loop gain is typically 200,000 (+106 dB). By itself, the open-loop gain is much too large to be used in most circuits. Such a large amount of gain tends to make these circuits unstable. However, we can decrease and stabilize the circuit gain of an op-amp stage and greatly enhance several other parameters that are very desirable for a linear amplifier by adding a few external components. This process is known as adding *negative feedback*. Like most things in life, you do not get something for nothing. Here, the tradeoff is a loss in circuit gain for an increase in stability.

To apply negative feedback, a fraction of the output signal that is 180° out of phase with the incoming signal is added to the incoming signal. The feedback network usually consists of passive components such as resistors and capacitors.

The voltage gain of the circuit with a closed feedback path between the output and the amplifier's input is referred to as the *closed-loop voltage gain* (A_{CL}). If no feedback is used, the circuit gain would then equal the amplifier's open-loop gain. The open-loop gain and the closed-loop gain are interrelated by what is referred to as the *loop gain* (A_L), so that

$$A_L = \frac{A_{OL}}{A_{CL}} \qquad \text{(Eq. 1-2)}$$

THE INVERTING OPERATIONAL AMPLIFIER

One way to implement negative feedback with an op-amp is to have the feedback network connected parallel to the amplifier's input and output, as shown in Figure 1-3. Here, the feedback network consists of a single resistor (R_2), called the *feedback resistor*, while resistor R_1 is frequently called the *input resistor*. This circuit is called an *inverting amplifier* as its output is inverted (i.e., of opposite polarity) with respect to its input signal. Also, notice that the input signal is connected to the op-amp's inverting input through the input resistor R_1.

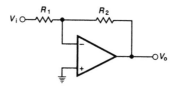

Fig. 1-3. Inverting amplifier with a closed-loop gain of $-R_2/R_1$.

The closed-loop voltage for the inverting amplifier is

$$A_{CL} = \frac{V_o}{V_i} \qquad \text{(Eq. 1-3a)}$$

$$A_{CL} = -\frac{R_2}{R_1} \qquad \text{(Eq. 1-3b)}$$

The output voltage is then

$$V_o = -\left(\frac{R_2}{R_1}\right)V_i \qquad \text{(Eq. 1-4)}$$

The minus sign in Equation 1-4 indicates that the output voltage's polarity is opposite that of the input voltage. When the polarity of the input signal is positive, the output signal is negative, and vice versa. For AC signals, this represents a 180° phase shift. Closer inspection of Equation 1-3b reveals that the closed-loop gain can be less than 1 (R_1 larger than R_2), equal to 1 (R_1 equal to R_2), or greater than 1 (R_2 greater than R_1).

For the ideal inverting amplifier circuit there are no restrictions to what values R_1 and R_2 may have because the closed loop-gain is based only on the ratio of their values. However, there are several practical considerations that should be kept in mind and are discussed in Chapter 2.

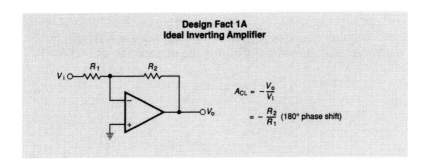

EXAMPLE 1-1

Figure 1-4 shows the schematic of an inverting amplifier having a DC input voltage of +0.5 V. Resistor R_2 (20 kΩ) is made 20 times larger than R_1 (1 kΩ) for a closed-loop gain of 20. Also, the output voltage will be the opposite polarity and 20 times larger than +0.5 V input or −10 V.

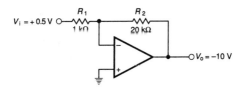

Fig. 1-4. Inverting amplifier with a closed-loop gain of 20.

THE NONINVERTING OPERATIONAL AMPLIFIER

Another way of implementing negative feedback with an op-amp is shown in Figure 1-5. The feedback section is a simple resistance voltage divider consisting of R_1 and R_2 connected in series. Here, R_2 is the *feedback resistor*, while R_1 is the *input resistor*. This circuit is called a *noninverting amplifier* because its output is always the

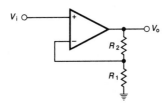

Fig. 1-5. Noninverting amplifier having closed-loop voltage gain of 1 + R_2/R_1.

same polarity as its input signal. In addition, notice that the input signal is connected directly to the op-amp's noninverting input. The closed-loop voltage gain for the noninverting amplifier is

$$A_{CL} = \frac{V_o}{V_i}$$ (Eq. 1-5a)

$$A_{CL} = 1 + \frac{R_2}{R_1}$$ (Eq. 1-5b)

The output voltage is then

$$V_o = \left(1 + \frac{R_2}{R_1}\right) V_i$$ (Eq. 1-6)

and the output voltage will always be greater than the input voltage. Also, since the input signal is applied to the op-amp's noninverting input, the output voltage is always in phase with the input for AC signals.

As with the ideal inverting amplifier circuit, there are no restrictions to what values R_1 and R_2 may have because the closed loop-gain is based only on the ratio of their values. However, there are several practical considerations that should be kept in mind and are discussed in Chapter 2.

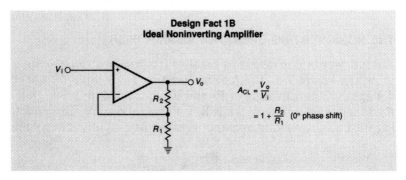

Design Fact 1B
Ideal Noninverting Amplifier

EXAMPLE 1-2

Figure 1-6 shows the schematic of a noninverting amplifier having a closed-loop gain of 40 and a DC input voltage of +10 mV. Resistor

R_2 (39 kΩ) must be then made 39 times larger than R_1 (1 kΩ). Also, the output voltage will be the same polarity and 40 times larger than +10 mV, or +0.4 V.

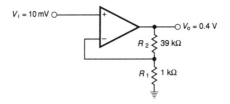

Fig. 1-6. Noninverting amplifier with a closed-loop gain of 40.

THE VOLTAGE FOLLOWER

If for the noninverting amplifier circuit of Figure 1-5 we short-circuit R_2 ($R_2 = 0$) and open-circuit R_1 ($R_1 = \infty$), we get a very simple but useful circuit. It is called a *voltage follower* and is shown in Figure 1-7. In this case, Equation 1-5 then simplifies to the output voltage being exactly equal to the input voltage, such that the output signal *follows* its input. In addition, the output and input signals are exactly in phase (i.e., same polarity).

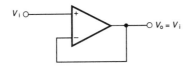

Fig. 1-7. Op-amp voltage follower having unity closed loop gain.

In comparison with an ideal amplifier, the voltage follower has essentially an infinite input impedance and zero output impedance. It then serves as a *buffer amplifier*. The output equals the input, in both signal amplitude and polarity (or 0° phase shift).

SUMMING AMPLIFIERS

If several input resistors are added to the inverting input of the inverting amplifier of Figure 1-3, the result is what is known as a *summing amplifier*, or *adder* (Fig. 1-8). The output of the summing amplifier is proportional to the *algebraic sum* of its separate inputs. When all input signals are of the audio type (20 Hz–20 kHz), the summing amplifier is frequently called a *signal mixer*. The mixer

9

then is used to combine audio signals from several microphones, guitars, tape recorders, etc., to provide a single output.

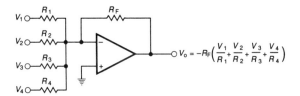

Fig. 1-8. A four-input inverting summing amplifier that amplifies the algebraic sum of its inputs.

Each input signal sees a closed-loop gain dependant on the single feedback resistor (R_F) and the corresponding input resistor (e.g., R_F/R_1, R_F/R_2, etc.). The output voltages due to each input signal are then added, so that for a four-input summing amplifier

$$V_o = -\left(\frac{R_F}{R_1}\right)V_1 - \left(\frac{R_F}{R_2}\right)V_2 - \left(\frac{R_F}{R_3}\right)V_3 - \left(\frac{R_F}{R_4}\right)V_4 \quad \text{(Eq. 1-7a)}$$

or

$$V_o = -R_F\left(\frac{V_1}{R_1} + \frac{V_2}{R_2} + \frac{V_3}{R_3} + \frac{V_4}{R_4}\right) \quad \text{(Eq. 1-7b)}$$

The polarity of the resultant output voltage will be the opposite of the algebraic sum of the various input signals.

For the special case where all four input resistors are made equal to the feedback resistor ($R_1 = R_2 = R_3 = R_4 = R_F$), Equation 1-7b simplifies to

$$V_o = -(V_1 + V_2 + V_3 + V_4) \quad \text{(Eq. 1-8)}$$

which is the inverted algebraic sum of all the inputs. The input signals can be all DC voltages, all AC voltages of the same or different waveforms and frequencies, or a mixture of AC and DC signals. The output signal will nevertheless be proportional to the *instantaneous algebraic sum* of all the input signals.

EXAMPLE 1-3

The following three DC voltages serve as inputs to the summing amplifier of Figure 1-9:

$$V_1 = +2V$$
$$V_2 = -1V$$
$$V_3 = -3V$$

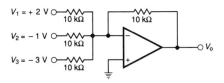

Fig. 1-9. Summing amplifier circuit.

Since all resistors are equal, the output voltage is then the inverted algebraic sum of all three inputs, or

$$V_o = -[2V + (-1V) + (-3V)]$$
$$= +2V$$

One nice variation of the summing amplifier circuit is *the averager*, which takes the instantaneous algebraic average of all its inputs. To achieve this, all input resistors are made the same (i.e., $R_1 = R_2 = R_3$, etc.), while the feedback resistor is made equal to

$$R_F = \frac{R_1}{N} \qquad \text{(Eq. 1-9)}$$

where N is the number of input signals to be averaged. For the four-input summing amplifier of Figure 1-8, the output voltage will represent the instantaneous average of its four inputs when all four input resistors are equal and $R_F = R_1/4$.

EXAMPLE 1-4

Determine the component values for an inverting summing amplifier that will take the average of two input signals.

Assuming that both input resistors are made equal to a given standard value (such as 20 kΩ), the feedback resistor must be equal to 20 kΩ/2 or 10 kΩ, giving the final circuit of Figure 1-10.

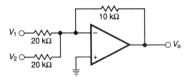

Fig. 1-10. Average circuit example.

In addition to the inverting summing amplifier, it is possible to have a *noninverting* summing amplifier, as shown by the circuit of Figure 1-11. In actuality, it is nothing more than the process of passively summing the voltages at a single node followed by a conventional, noninverting amplifier. If all the resistors that are in series with each of the N inputs are made equal for convenience to R_1, then the feedback resistor R_F must equal

$$R_F = (N - 1)R_1 \qquad \text{(Eq. 1-10)}$$

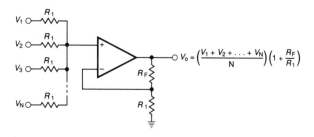

Fig. 1-11. Noninverting summing amplifier.

EXAMPLE 1-5

Determine the component values for a noninverting summing amplifier that will take the average of three input signals.

Assuming that all three input resistors are made equal to a given standard value of 10 kΩ, the feedback resistor must be made equal to

12

$$R_F = (3-1)(10 \text{ k}\Omega)$$
$$= 20 \text{ k}\Omega$$

(Eq. 1-10)

as shown in Figure 1-12.

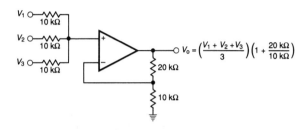

Fig. 1-12. Completed example circuit for a summing amplifier.

THE DIFFERENCE AMPLIFIER

As shown in Figure 1-13a, a difference, or differential amplifier has input voltages that are applied simultaneously to both the inverting and noninverting inputs. Its output voltage is proportional to the voltage difference ($V_2 - V_1$).

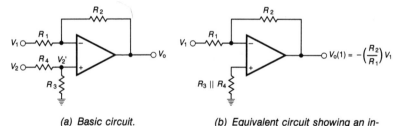

(a) Basic circuit.

(b) Equivalent circuit showing an inverting amplifier by reducing V_2 to zero.

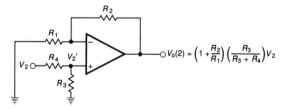

(c) Equivalent circuit showing a noninverting amplifier with voltage divider (R_3–R_4) by reducing V_1 to zero.

Fig. 1-13. Difference amplifier that amplifies the input voltage difference $V_2 - V_1$.

Although this circuit looks complex, the analysis is really quite simple using the principles of inverting and noninverting amplifiers. If V_2 is first set to zero (Fig. 1-13b), we have an inverting amplifier with a closed-loop gain equal to $-R_2/R_1$. The output voltage is then

$$V_o(1) = -\left(\frac{R_2}{R_1}\right)V_1 \qquad \text{(Eq. 1-11)}$$

when $V_2 = 0$

Next, remove the short circuit across V_2 and set only V_1 equal to zero. The circuit now resembles a noninverting amplifier with a closed-loop gain of $1 + R_2/R_1$ preceded by a voltage divider composed of R_3 and R_4 (Fig. 1-13c). The actual voltage that the op-amp sees is V_2' which is related to the input voltage V_2 by the voltage divider relation

$$(V_2)' = \frac{R_3}{R_3 + R_4}V_2 \qquad \text{(Eq. 1-12)}$$

when $V_1 = 0$

The noninverting output is then

$$V_o(2) = \left(1 + \frac{R_2}{R_1}\right)(V_2)' \qquad \text{(Eq. 1-13a)}$$

or

$$V_o(2) = \left(1 + \frac{R_2}{R_1}\right)\left(\frac{R_3}{R_3 + R_4}\right)V_2 \qquad \text{(Eq. 1-13b)}$$

The total output signal of the difference amplifier from both input signal sources is then the sum of Equations 1-11 and 1-13b

$$V_o = V_o(1) + V_o(2)$$

$$= \left(1 + \frac{R_2}{R_1}\right)\left(\frac{R_3}{R_3 + R_4}\right)V_2 - \left(\frac{R_2}{R_1}\right)V_1 \qquad \text{(Eq. 1-14)}$$

The output voltage then depends on the values of four resistors. To simplify matters, two pairs are chosen so that

$$R_1 = R_4$$
$$R_2 = R_3$$

With these simplifications, Equation 1-14 reduces to

$$V_o = \frac{R_2}{R_1}(V_2 - V_1) \qquad \text{(Eq. 1-15)}$$

Here, the resistance ratio R_2/R_1 is referred to as the *differential gain* (A_D) so that the output voltage is the product of the differential gain and the differential input voltage, $V_2 - V_1$.

When all four resistors are made equal, Equation 1-15 reduces to

$$V_o = V_2 - V_1 \qquad \text{(Eq. 1-16)}$$

Such a circuit is then called a unity-gain *analog subtractor*.

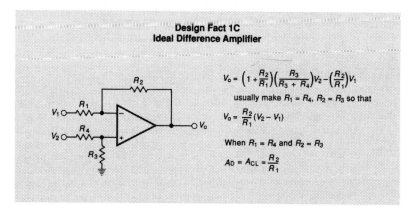

Design Fact 1C
Ideal Difference Amplifier

$V_o = \left(1 + \frac{R_2}{R_1}\right)\left(\frac{R_3}{R_3 + R_4}\right)V_2 - \left(\frac{R_2}{R_1}\right)V_1$

usually make $R_1 = R_4$, $R_2 = R_3$ so that

$V_o = \frac{R_2}{R_1}(V_2 - V_1)$

When $R_1 = R_4$ and $R_2 = R_3$

$A_D = A_{CL} = \frac{R_2}{R_1}$

EXAMPLE 1-6

Determine the component values for a difference amplifier having a differential gain of 150.

For simplicity choose

$$R_1 = R_4$$
$$R_2 = R_3$$

If R_1 is chosen as 1 kΩ, then R_2 must be 150 kΩ for a differential gain of 150 (Eq. 1-15), as shown in Figure 1-14.

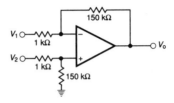

Fig. 1-14. Completed example circuit for a difference amplifier.

THE INSTRUMENTATION AMPLIFIER

It is often desired to be able to vary the gain of a difference amplifier using a single potentiometer. For the basic difference amplifier circuit of Figure 1-13, the ratios R_2/R_1 and R_4/R_2 must always be equal. If R_2 is increased by 25%, then R_4 also must be increased by 25%. As a common practice, it is desired to change the performance of a given circuit by varying the value of only one component, rather than simultaneously changing two or more values.

An improvement of the basic difference amplifier is the *instrumentation amplifier*, or *IA* (Fig. 1-15). It adds two op-amps (A_1 and A_2) and three resistors in front of the basic difference amplifier connection. Amplifiers A_1 and A_2 are wired as voltage followers and function as high-impedance buffer amplifiers for the two input voltages, V_1 and V_2. Except for R_G, all the remaining six resistors are made the same value. For this arrangement, the differential gain is then

$$A_D = 1 + \frac{2R_1}{R_G} \qquad \text{(Eq.1-17)}$$

The output voltage of the ideal instrumentation amplifier is its differential gain times the differential input voltage

$$V_o = \left(1 + \frac{2R_1}{R_G}\right)(V_2 - V_1) \qquad \text{(Eq. 1-18)}$$

Like the other six resistors, the gain-setting resistor R_G could be a fixed value so that the differential gain would be fixed. However, by

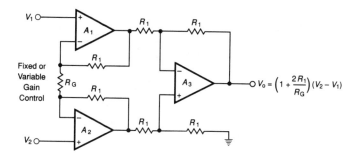

Fig. 1-15. Instrumentation amplifier with a single resistance gain control.

varying R_G alone, we are able to smoothly vary the differential gain. As R_G is decreased, A_D becomes larger. As R_G is increased towards infinity (an open circuit), A_D approaches unity. Instead of making an instrumentation amplifier from individual op-amps and resistors like the circuit of Figure 1-15, some manufacturers have produced an integrated circuit IA in a single package except for the gain-setting resistor R_G. In this case, the characteristics of the amplifier sections are matched, as are the six internal resistors (R_1).

EXAMPLE 1-7

The differential gain of the instrumentation amplifier circuit of Figure 1-16 can be easily varied as the 100-kΩ potentiometer varies over its entire range.

Resistance R_G is the series combination of the 100-Ω resistor and the 100-kΩ potentiometer. When the potentiometer resistance is zero, then $R_G = 100\ \Omega$, and the differential gain is

$$A_D = 1 + \frac{(2)(100\ k\Omega)}{100\ \Omega}$$

$$= 2001$$

(Eq. 1-17)

When the potentiometer resistance is at its maximum value, then $R_G = 100.1$ kΩ, and the differential gain is

$$A_D = 1 + \frac{(2)(100\ k\Omega)}{100.1\ k\Omega}$$

$$= 2.998$$

(Eq. 1-17)

17

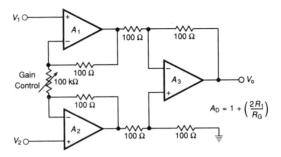

Fig. 1-16. Circuit example of an instrumentation amplifier circuit.

For all practical purposes, the differential gain for the circuit of Figure 1-16 would vary from 3 to 2000.

One common application for either the IA or difference amplifier is amplifying the output of a bridge circuit, such as the Wheatstone resistance bridge of Figure 1-17. The output of the bridge is the differential voltage $V_2 - V_1$, where V_1 is the voltage at node 1 with respect to ground and V_2 is the voltage at node 2 with respect to ground.

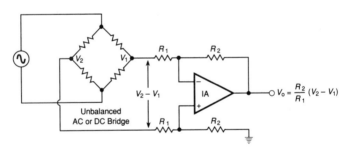

Fig. 1-17. Using a difference amplifier with a Wheatstone resistance bridge to convert a differential floating input into a single-ended output.

Another application is one that allows an oscilloscope to have a single-ended input to display the voltage between two circuit nodes, neither of which are at ground. As shown in Figure 1-18, a unity-gain difference amplifier is used to enable the oscilloscope to display the AC voltage across the capacitor. Since neither end of the capacitor is grounded, the oscilloscope cannot be connected to either end of the capacitor without damaging the circuit or giving erroneous readings.

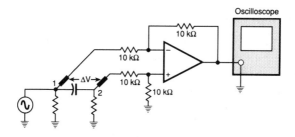

Fig. 1-18. Using a difference amplifier to enable a single-ended oscilloscope to measure a differential voltage.

JFET-INPUT OPERATIONAL AMPLIFIERS

Besides the standard bipolar-type op-amp, there is another category of op-amp devices called the *JFET* (junction field effect transistor) op-amp. These devices (such as BIFETs*) have a JFET input stage for both the inverting and noninverting inputs instead of ordinary equivalent bipolar transistors. As a result, JFET-input op amps require low-supply currents, yet maintain a wide frequency response and high-switching speed. Well-matched, high-voltage JFET-input devices also provide very low input bias and offset currents. In virtually all cases, JFET devices do not require any circuit change when used in place of standard bipolar op-amps.

One example of a BIFET is the LF351, and it is pin-for-pin compatible with a standard 741 device. This allows users to upgrade the overall performance of existing circuits using 741 op-amps merely by substituting the LF351 for the 741.

CMOS OPERATIONAL AMPLIFIERS

CMOS is the acronym for *Complimentary Metal Oxide Semiconductor*. It is constructed using both p- and n-channel (i.e, *complimentary*) MOS transistors whose characteristics are matched.

Compared with conventional bipolar operation amplifiers, most CMOS op-amps provide almost an infinite input impedance and an output that can swing from one supply rail to the other, a feature not available previously. Because CMOS devices require very little

---

* The term *BIFET* is a registered trademark of National Semiconductor.

power, the CMOS op-amp is ideally suited for battery-powered or energy-saving applications.

DEVICE IDENTIFICATION

Like other semiconductor devices such as transistors and diodes, the particular part number of an op-amp device generally has no correlation to why it is different from another. For example, there is no technical information that can be gained from the number 741 (an op-amp) that would tell us why it is different than either a 318 (another op-amp) or a 309 (a voltage regulator). The numbers for op-amp devices (as well as all integrated circuit devices) were assigned by their original manufacturers, either in some logical sequence known only to them, or in a haphazard, random fashion that makes no sense to the user.

Despite the confusion that may occur to the first-time user, there are several aspects about the device identification system that are somewhat straightforward. For a given identification code marked on the device, it is usually possible to determine the manufacturer, product grade, package material, and date of manufacture, all which try to answer the questions: who, what, how good, and when.

Manufacturer Prefixes

In all probability, a given device (such as the popular type 741 op amp) will be made by more than one manufacturer. To identify the manufacturer, a prefix is often added to the device number marked on the package. As examples, a type-741 op-amp may be marked in one of the following ways: µA741, SN741, NE741, CA741, MC1741, or RC741. The most commonly encountered manufacturer prefixes are listed in Table 1-1, noting that several prefixes are used by more than one manufacturer. To distinguish themselves, some manufacturers also include their distinctive *logo*, or corporate symbol on the device. For example, the logo for Texas Instruments is the outline of the state of Texas with the letters *TI*.

Grades

In most cases, a given op-amp is available in different grades. The top grade is for military or high-reliability applications. This is sometimes called a *mil-spec* device. On the other hand, the lowest grade is the *commercial* or *hobby-grade* version. The offset parameters, allowable operating and storage temperature ranges for the mil-spec version, are usually better. Of course these improvements do not come without their price since mil-spec graded devices are

Table 1-1. Manufacturer Identification Prefix Codes

Manufacturer	Prefixes
Analog Devices	AD
Burr-Brown	BB
Exar	XR
Fairchild	NE, µA
Harris Semiconductor	HA
Intersil	ICL, ICM, NE, SE
Lithic Systems	LC
Motorola	MC
National Semiconductor	LF, LM, LP
Precision Monolithics	OP
Raytheon	RC, RM
RCA	CA, CD
Signetics	NE, SE, µA
Silicon General	SG
Texas Instruments	SN

more expensive than their commercial/hobby-grade counterparts. Typically, this premium may be approximately five times the cost of the commercial device.

To distinguish the various performance grades, some manufacturers use distinctive prefixes, others use appropriate suffixes, while still others use a number code. For example, Raytheon uses the *RC* and *RM* prefixes respectively to identify its commercial (e.g., RA741) and military (RM741) versions. The *SE/NE* prefixes used by some manufacturers respectively refer to the military (e.g., SE5558) and commercial/hobby-grade (NE5558) versions.

On the other hand, some manufacturers (such as Signetics) use the letter *C* as a suffix to indicate its commercial-grade device. A µA741C is the commercial-grade type-741 op-amp, while the µA741 is its designation for the military-grade 741 device. Still further, National Semiconductor designates many of their devices in terms of a 100, 200, or 300 series. These three series refer to the military, commercial, and hobby-grade specifications respectively. For example, the same op-amp manufactured by National Semiconductor may be designated as either an LM118, LM218, or LM318. As a general rule, the 300-series hobby-grade number is generally used as the generic number to describe a particular device. Most users would say, "use a 318 op-amp" to mean that either an LM118, LM218, or LM318 could be used as appropriate.

21

Package Material Suffixes

To accurately complete the description of a particular device, a suffix code letter is added to signify the type of material the package is made from. The most common letters are C (for ceramic) and P (for plastic). Other codes are usually found in the manufacturer's data sheet. In most circumstances however, these codes are of minor importance.

Date of Manufacture

Analog devices, and to a greater extent, digital devices, are marked to indicate their date of manufacture. Very simply, the date is represented by a four-digit code representing the year and number of the week. For example, the code *8521* translates as manufactured during the 21st week in 1985. Since the week number cannot be greater than 52 (or possibly 53 in certain calendar years), it is then impossible to get the year and week codes confused. When we reach the year 2000, perhaps a new method will be devised to eliminate confusion. Other than curiosity, the date of manufacture is not that important.

PACKAGE STYLES

Op-amps are available in several package styles. Of these, the most popular is referred to as the *DIP*, which is the acronym for *dual-in-line package*. The DIP configuration always has an even number of pins spaced 0.1 inch apart in two equal rows, which make them perfect for use with solderless breadboarding sockets as well as printed-circuit board layouts. Most op-amps are available in the 8-pin DIP package, although some come in either 14 or 16-pin configurations. Since the 8-pin DIP is the smallest, it is often referred to as a "mini-DIP."

Using the 8-pin DIP of Figure 1-19a as a guide, the pins are numbered in a counterclockwise fashion starting with the number 1 pin located in the upper left-hand corner. It is very important that the number 1 pin can be correctly identified because the device can be easily destroyed if the power supply is connected to the wrong terminals. One end of the package is usually different. It has either a semicircular notch or channel or a *dot* in one corner. If this end becomes designated as the *top*, then the number 1 pin is the top pin of the left-hand row of pins. This holds true no matter how many pins there are in a DIP.

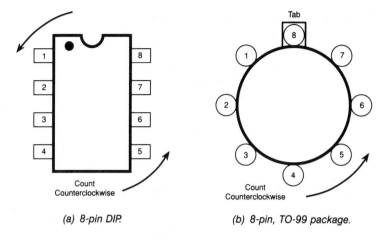

Fig. 1-19. Determining pin numbers.

Besides the DIP configuration, op-amps (to a lesser extent) are available in various TO package styles. The abbreviation *TO* stands for *transistor outline*, and forms the basis of a numbering system created by the Joint Electron Device Engineering Council (JEDEC), an industry association. For op-amps, these body styles are designated by numbers such as T0-99 (8 leads), while the standard 14-pin DIP is also referred to as a TO-116 style package.

The 8-pin, TO-99 style metal can is shown in Figure 1-19b. These devices generally look like transistors, except they have more leads which are arranged in a cylindrical formation. The highest pin number is the one located underneath the tab, as viewed from the top of the case. Like the DIP configuration, the pins are then numbered in a counterclockwise fashion starting with the number 1 pin located immediately to the left of tab.

SOME PRECAUTIONS

As with any type circuit, certain rules should be followed so that a circuit will operate as it was intended to do, as well as preventing possible destruction of the circuit or components.

- Always wire circuits or make component/wiring changes with the power off.
- Connect signal sources *after* DC power has been applied to the circuit. Conversely, disconnect signal sources before removing or turning off DC power from the circuit.

- Make sure that the power-supply polarities to the op-amp device are correct. In general, these connections should be made first when breadboarding so they will not be forgotten later.
- Many times an op-amp circuit will either oscillate, or the noise originating from the power line or adjacent signals will appear at its output. Generally, a 0.1-µF bypass capacitor should be connected between the power supply pins of the device and ground. For devices using split supplies, bypass capacitors are connected between the positive supply pin and ground as well as the negative supply pin and ground. For those devices powered from a single supply voltage, the bypass capacitor is connected only between the positive supply pin and ground. However, it should be strongly emphasized that these connections should be made as close as possible to the supply pins of the device(s) and its ground pin. Furthermore, all connections should be made as short as possible for optimum effectiveness in decoupling extraneous signals from the DC power supply lines.

Chapter 2

Practical Performance Considerations

INTRODUCTION AND OBJECTIVES

The concepts of the ideal op-amp developed in Chapter 1 are often used to design many basic op-amp circuits. In applications where performance requirements are not very stringent, this approach is satisfactory. For more demanding applications however, many of the actual characteristics of the op-amp must be taken into consideration if the circuit is to meet given specifications.

At the completion of this chapter, you will be able to:

- Read and understand a data sheet of a typical op-amp.
- Determine the magnitude of output offset voltage as a result of an op-amp's input offset voltage, input bias current, and input offset current.
- Explain how to externally compensate for output offset voltages.
- Determine the input and output resistances of amplifier stages.
- Explain how the frequency response of most op-amp circuits is affected by the gain-bandwidth product, rise time, and slew rate.
- Calculate the effect of the common-mode rejection on the performance of a difference amplifier.

THE DATA SHEET

Perhaps the best way to understand the many factors that may affect the performance of an op-amp is to first examine its data sheet. Figure 2-1 shows the data sheet of the type-741 op-amp, which is perhaps the most popular general-purpose type in use. Although

HIGH PERFORMANCE OPERATIONAL AMPLIFIER µA741

LINEAR INTEGRATED CIRCUITS

DESCRIPTION
The µA741 is a high performance operational amplifier with high open loop gain, internal compensation, high common mode range and exceptional temperature stability. The µA741 is short-circuit protected and allows for nulling of offset voltage.

FEATURES
- INTERNAL FREQUENCY COMPENSATION
- SHORT CIRCUIT PROTECTION
- OFFSET VOLTAGE NULL CAPABILITY
- EXCELLENT TEMPERATURE STABILITY
- HIGH INPUT VOLTAGE RANGE
- NO LATCH-UP

ABSOLUTE MAXIMUM RATINGS

	µA741C	µA741
Supply Voltage	±18V	±22V
Internal Power Dissipation (Note 1)	500mW	500mW
Differential Input Voltage	±30V	±30V
Input Voltage (Note 2)	±15V	±15V
Voltage between Offset Null and V⁻	±0.5V	±0.5V
Operating Temperature Range	0°C to +70°C	-55°C to +125°C
Storage Temperature Range	-65°C to +150°C	-65°C to +150°C
Lead Temperature (Solder, 60 sec)	300°C	300°C
Output Short Circuit Duration (Note 3)	Indefinite	Indefinite

Notes
1. Rating applies for case temperatures to 125°C; derate linearly at 6.5mW/°C for ambient temperatures above +75°C.
2. For supply voltages less than ±15V, the absolute maximum input voltage is equal to the supply voltage.
3. Short circuit may be to ground or either supply. Rating applies to +125°C case temperature or +75°C ambient temperature.

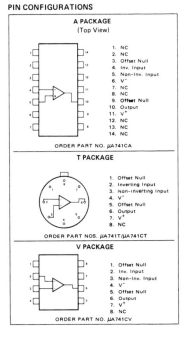

EQUIVALENT CIRCUIT

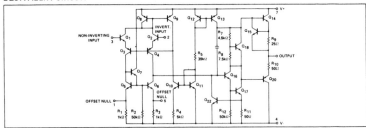

Fig. 2-1. Manufacturer's data sheet for the µA741 op-amp.
(Courtesy of Signetics Corporation)

this particular data sheet is for the unit made by Signetics, type-741 op-amps made by other manufacturers, such as RCA and National Semiconductor, will have nearly the same specifications. The data sheet for the op-amp (as well as most other types of integrated circuits) usually contains the following information:

µA741 – HIGH PERFORMANCE OPERATIONAL AMPLIFIER

ELECTRICAL CHARACTERISTICS ($V_S = \pm 15V$, $T_A = 25°C$ unless otherwise specified)

PARAMETER	MIN.	TYP.	MAX.	UNITS	TEST CONDITIONS
		µA741C			
Input Offset Voltage		2.0	6.0	mV	$R_S \leq 10k\Omega$
Input Offset Current		20	200	nA	
Input Bias Current		80	500	nA	
Input Resistance	0.3	2.0		MΩ	
Input Capacitance		1.4		pF	
Offset Voltage Adjustment Range		±15		mV	
Input Voltage Range	±12	±13		V	
Common Mode Rejection Ratio	70	90		dB	$R_S \leq 10k\Omega$
Supply Voltage Rejection Ratio		10	150	µV/V	$R_S \leq 10k\Omega$
Large-Signal Voltage Gain	20,000	200,000			$R_L \geq 2k\Omega$, $V_{out} = \pm 10V$
Output Voltage Swing	±12	±14		V	$R_L \geq 10k\Omega$
	±10	±13		V	$R_L \geq 2k\Omega$
Output Resistance		75		Ω	
Output Short-Circuit Current		25		mA	
Supply Current		1.4	2.8	mA	
Power Consumption		50	85	mW	
Transient Response (unity gain)					$V_{in} = 20mV$, $R_L = 2k\Omega$, $C_L \leq 100pF$
Risetime		0.3		µs	
Overshoot		5.0		%	
Slew Rate		0.5		V/µs	$R_L \geq 2k\Omega$
The following specifications apply for $0°C \leq T_A \leq +70°C$					
Input Offset Voltage			7.5	mV	
Input Offset Current			300	nA	
Input Bias Current			800	nA	
Large Signal Voltage Gain	15,000				$R_L \geq 2k\Omega$, $V_{out} = \pm 10V$
Output Voltage Swing	±10	±13		V	$R_L \geq 2k\Omega$
		µA741			
Input Offset Voltage		1.0	5.0	mV	$R_S \leq 10k\Omega$
Input Offset Current		10	200	nA	
Input Bias Current		80	500	nA	
Input Resistance	0.3	2.0		MΩ	
Input Capacitance		1.4		pF	
Offset Voltage Adjustment Range		±15		mV	
Large-Signal Voltage Gain	50,000	200,000			$R_L \geq 2k\Omega$, $V_{out} = \pm 10V$
Output Resistance		75		Ω	
Output Short Circuit Current		25		mA	
Supply Current		1.4	2.8	mA	
Power Consumption		50	85	mW	
Transient Response (unity gain)					$V_{in} = 20mV$, $R_L = 2k\Omega$, $C_L \leq 100pF$
Risetime		0.3		µs	
Overshoot		5.0		%	
Slew Rate		0.5		V/µs	$R_L \geq 2k\Omega$
The following specifications apply for $-55°C \leq T_A \leq +125°C$					
Input Offset Voltage		1.0	6.0	mV	$R_S \leq 10k\Omega$
Input Offset Current		7.0	200	nA	$T_A = +125°C$
		20	500	nA	$T_A = -55°C$
Input Bias Current		0.03	0.5	µA	$T_A = +125°C$
		0.3	1.5	µA	$T_A = -55°C$
Input Voltage Range	±12	±13		V	
Common Mode Rejection Ratio	70	90		dB	$R_S \leq 10k\Omega$
Supply Voltage Refection Ratio		10	150	µV/V	$R_S \leq 10k\Omega$
Large-Signal Voltage Gain	25,000				$R_L \geq 2k\Omega$, $V_{out} = \pm 10V$
Output Voltage Swing	±12	±14		V	$R_L \geq 10k\Omega$
	±10	±13		V	$R_L \geq 2k\Omega$
Supply Current		1.5	2.5	mA	$T_A = +125°C$
		2.0	3.3	mA	$T_A = -55°C$
Power Consumption		45	75	mW	$T_A = +125°C$
		45	100	mW	$T_A = -55°C$

Fig. 2-1. (Cont.)

- A general description of the op-amp.
- An internal equivalent circuit schematic.
- Pin configuration and packaging styles.
- Absolute maximum ratings.
- Electrical characteristics.

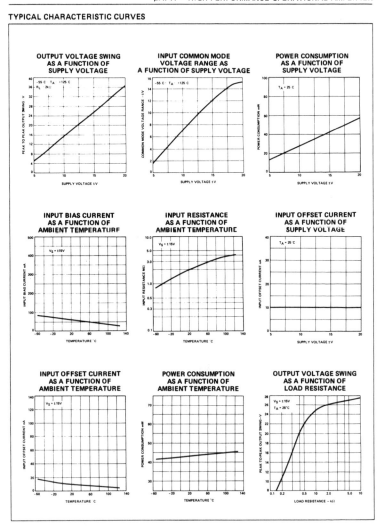

Fig. 2-1. (Cont.)

- Typical performance curves.
- Typical application circuits and design information.

Most of the important parameters are discussed in this section, using the type-741 op-amp as a representative example. Later sections in this chapter discuss how these parameters then affect circuit perfor-

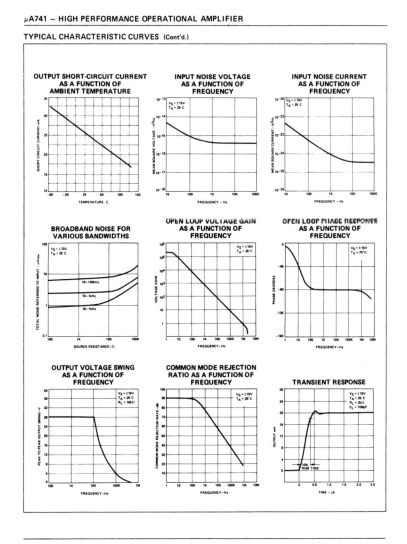

Fig. 2-1. (Cont.)

mance. For example, Signetics manufactures both a 741 (or military-grade) and 741C (or commercial-grade version). Unless specified otherwise, the commercial-grade (741C) is assumed whenever the parameters of the 741-type device are discussed.

General Description

The general description can provide a number of interesting insights into the features of the particular device. There may be an indication as to whether or not the type of op-amp has FET input stages, is internally compensated, or has information about those characteristics that may lead to improved performance.

Internal Equivalent Circuit

Although the internal diagram will have little impact on the way the op-amp is ultimately used, it does demonstrate the intricacies of the internal circuit. One can readily appreciate the effort needed to design and fabricate even the simplest of integrated circuit devices. For the purposes of this book, however, the internal circuit of the op-amp will not be addressed directly.

Maximum Ratings

The maximum ratings listed are the maximum values that the op-amp can safely handle without the possibility of destruction. Under no circumstances should they be equalled or exceeded.

1. **Supply Voltage ($+V_{CC}$, $-V_{EE}$)**

This is the maximum positive and negative supply voltage that can be used to power the op-amp.

2. **Internal Power Dissipation (P_D)**

This is the maximum power that the op-amp is capable of dissipating at a specified ambient temperature. For higher temperatures, this must be *derated*, or reduced accordingly.

3. **Differential Input Voltage (V_{id})**

This is the maximum voltage that can be applied between the inverting and noninverting inputs.

4. **Input Voltage (V_{icm})**

Also called the *common-mode* voltage, this is the maximum input voltage than can be simultaneously applied between both inputs and ground. In general, this is equal to the positive supply voltage.

5. Operating Temperature (T_A)

This is the ambient temperature for which the op-amp will operate within the manufacturer's specifications. The military-grade version usually has a wider temperature range than the commercial version.

6. Output Short-Circuit Duration

This is the time duration that the op-amp's output terminal can be short-circuited to ground or either supply voltage without damage. For the 741, the output can be short circuited indefinitely.

Electrical Characteristics

The electrical characteristics are those parameters that can severely affect or limit the performance of many op-amp circuits. Those cited are usually specified for a given supply voltage and operating temperature unless specified otherwise. For the 741 and 741C versions listed in the data sheet, the specifications are given for supply voltages of $V_{CC} = +15$ V, $V_{EE} = -15$ V, and an ambient operating temperature of 25°C. In addition, other conditions may apply, such as a specified load resistance. Each parameter listed may have a minimum, typical, and/or maximum value. Electrical parameters can apply to static (DC) and dynamic (AC) circuit operation.

Static Input Parameters

1. Input Offset Voltage (V_{io})

This is the equivalent DC voltage that must be applied to one of the input terminals to produce a zero output voltage with the other input grounded. For the ideal op-amp, the output offset voltage is zero. For the 741, however, the input offset voltage is typically 2 mV and can be of either polarity.

2. Input Bias Current (I_B)

This is the average of the two DC currents that flow into the inverting and noninverting inputs. For an ideal op-amp, input bias current is zero due to its infinite impedance. For the 741, the input bias current is typically 80 nA.

3. Input Offset Current (I_{os})

This is the difference of the two input bias currents, whose polarity can be either positive or negative. For the 741, the input offset current is typically 20 nA.

4. Input Resistance (R_i)

This is the internal input resistance (or impedance) as seen at either the inverting or noninverting input to ground while the remaining input terminal is grounded. For an ideal device, it is infinite. For the 741, it is typically 2 MΩ.

Static Output Parameters

1. Output Resistance (R_{oi})

This is the internal output resistance (or impedance) as seen from the output terminal to ground. For an ideal op-amp, it is zero. For the 741, it is typically 75 Ω.

2. Output Short-Circuit Current (I_{osc})

This is the maximum DC output current that can be supplied to a load (e.g., 25 mA). This is specified when the output is short-circuited to ground.

3. Output Voltage Swing (V_{+SAT} and V_{-SAT})

Also called the *saturation voltage*, this is the maximum peak output voltage that the op-amp can produce without saturation or clipping. Its value is typically about two diode voltage drops (about 1.4 V) less than the corresponding supply voltages.

Dynamic Parameters

1. Open-Loop Gain (A_{OL})

Sometimes called the *large signal gain*, this is the voltage gain of the op-amp without any external feedback and varies with frequency. As is briefly mentioned in Chapter 1, the DC open-loop gain for the 741 is typically 200,000 (+106 dB).

2. Slew Rate (SR)

This is the time rate of change of the output voltage with the op-amp circuit having a closed-loop gain of 1. Ideally, the output voltage should exactly follow the input signal without distortion. For a rapidly changing input signal, such as a square wave, the output should also resemble a square wave. However, the slew rate of an op-amp, as a result of the inability of its internal circuitry to drive capacitive loads, causes the output signal to increase (or decrease) at a slower rate than the corresponding input signal, as shown in Figure 2-2a. The slew rate is calculated by dividing the output voltage swing (ΔV_o) by the corresponding time interval (Δt)

$$SR = \frac{\Delta V_o}{\Delta t} \qquad \text{(Eq. 2-1)}$$

The effect of the slew rate on sine waves is shown in Figure 2-2b. For an ideal op-amp, the slew rate is infinite. For the 741, it is typically 0.5 V/μs

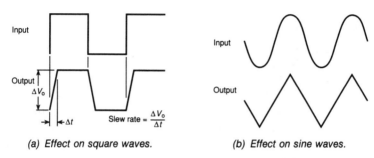

(a) Effect on square waves.   (b) Effect on sine waves.

Fig. 2-2. Slew rate in op-amps.

3. Unity-Gain Bandwidth (f_T)

This is the frequency at which the open-loop voltage gain decreases to 1 (0 dB). As is discussed later in this chapter, it is related to the op-amp's gain-bandwidth product.

4. Common-Mode Rejection Ratio (CMRR)

This is a measure of the ability of the op-amp to reject common-mode signals (i.e., signals that are simultaneously present at both inputs). Although the differential input voltage may be zero, there will exist a finite output as a result of the nonideal common-mode rejection.

The CMRR is a dimensionless ratio, but is usually expressed in decibels, which is then referred to as *common-mode rejection (CMR)*. The relation between CMRR and CMR is

$$CMR_{dB} = 20 \log_{10}(CMRR) \qquad \text{(Eq. 2-2)}$$

For the 741, the CMR is typically 90 dB, or a reduction of about 31,600 to 1.

5. Channel Separation

For packages having more than one internal op-amp, a certain amount of *crosstalk* will be present. That is, a signal applied to the input of one op-amp section will develop a small but finite output signal in the remaining section(s), *even though there is no input signal applied to the unused section(s)*. For the 747 and 5558 dual op-amps, the channel separation is typically 120 dB. For the 4156-quad (four) op-amp, it is 108 dB. This means that if one op-amp section of a 747 has an output of 1 V, there will be a signal that is 120 dB less (i.e., 1 µV) at the output of the unused section. For most applications however, channel separation is not a problem.

6. Transient Response, Rise Time (t_r)

This is the amount of time it takes for the output signal, in response to an input pulse, to rise from 10 to 90 percent of its steady-state value. For the 741, it is typically 0.3 µs.

The following sections now discuss how many of these parameters affect the performance of the ideal linear amplifier circuits discussed in Chapter 1. In addition, techniques are demonstrated that may minimize or eliminate some of the limitations imposed by the op-amp's parameters.

INPUT AND OUTPUT IMPEDANCES

For any op-amp circuit to be free from loading effects, the input impedance of the circuit should be high enough so that it does not load down the output of the signal source. At the same time, the op-amp's output impedance should be low enough so that any load that is connected to it does not load down the output of the circuit.

For the ideal op-amp, its input impedance is infinite while its output impedance is zero. From the data sheet of a typical op-amp, we now know that this is not the case. However, the application of negative feedback has several other advantages besides stabilizing the gain of the op-amp circuit.

Circuit Output Impedance

For voltage followers, inverting, noninverting, and difference amplifiers, negative feedback lowers the output impedance of the circuit. If the closed-loop gain is substantially smaller than the open-loop gain (which is usually the case), then the loop gain is much larger than 1. As a result, the closed-loop output impedance is approximately equal to

$$R_o(CL) \cong \frac{R_{oi}}{A_L} \qquad \text{(Eq. 2-3)}$$

As the loop gain is dependent on both the circuit's closed-loop gain and the op-amp's open-loop gain, Equation 2-3 can be rewritten as

$$R_o(CL) \cong \left(\frac{A_{CL}}{A_{OL}}\right) R_{oi} \qquad \text{(Eq. 2-4)}$$

For the case of a voltage follower whose closed-loop gain is 1, the output impedance of the circuit is approximately

$$R_o(CL) \cong \frac{R_{oi}}{A_{OL}} \qquad \text{(Eq. 2-5)}$$

When the loop gain is sufficiently large compared with the closed-loop gain, the output impedance of any linear amplifier circuit will then be substantially less than 1 Ω.

EXAMPLE 2-1

A noninverting amplifier circuit has a closed-loop voltage gain of 20 using a 741 op-amp with the following typical values obtained from its data sheet:

$$A_{OL} = 200{,}000 \; (+106 \; dB)$$
$$R_{oi} = 75\Omega$$

The output resistance of the noninverting amplifier is then

$$R_o(CL) = \left(\frac{20}{200{,}000}\right) 75\Omega \qquad \text{(Eq. 2-4)}$$
$$= 0.0075\Omega$$

For most circuits having negative feedback, the circuit output impedance calculation is not much to be concerned about. Even for open-loop circuits, such as a comparator (discussed in Chapter 6), the worst that the circuit's output impedance could be is equal to the op-amp's intrinsic output resistance, R_{oi}.

Circuit Input Impedance

The closed-loop input resistance of an op-amp circuit must be determined separately for inverting and noninverting amplifiers. For the inverting amplifier of Figure 2-3, the closed-loop input impedance is approximately equal to R_1

$$R_i(CL) \cong R_1 \qquad \text{(Eq. 2-6)}$$

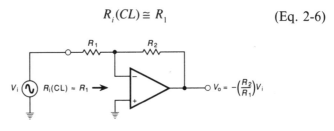

Fig. 2-3. Inverting amplifier input resistance.

Very often, the design of an inverting amplifier circuit requires that it have a specific input impedance. For a required closed-loop gain, the feedback resistor is then immediately determined as illustrated by the following example:

EXAMPLE 2-2

Determine the necessary component values for an inverting amplifier having an input impedance of 10 kΩ and a closed-loop voltage gain of 10 (+20 dB).

Since the required amplifier circuit input impedance is 10 kΩ, this immediately sets $R_1 = 10$ kΩ (Eq. 2-6). For a closed-loop voltage gain of 10, the feedback resistor (R_2) is then

$$R_2 = (10)(10 \ k\Omega)$$
$$= 100 \ k\Omega \qquad \text{(Eq. 1-3b)}$$

The completed circuit is shown in Figure 2-4.

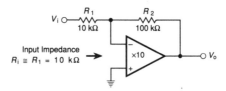

Fig. 2-4. Completed example circuit for an inverting amplifier.

For the noninverting amplifier circuit of Figure 2-5, the circuit's input impedance is significantly increased by the amplifier's loop gain, such that

$$R_i(CL) = (1 + A_L)R_i \qquad \text{(Eq. 2-7)}$$

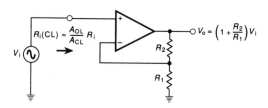

Fig. 2-5. Noninverting amplifier input resistance.

As the loop gain is usually much larger than 1, the circuit input impedance can be rewritten approximately in terms of the open- and closed-loop gains as

$$R_i(CL) \cong \left(\frac{A_{OL}}{A_{CL}}\right) R_i \qquad \text{(Eq. 2-8)}$$

In most cases, the effective closed-loop input impedance of the amplifier is usually so much greater than the op-amp's input impedance that it need not be determined as a practical matter. With the basic circuit of Figure 2-5, it is not possible to independently set the input impedance of a noninverting amplifier circuit.

CONTRIBUTIONS TO OUTPUT OFFSET VOLTAGE

The output voltage of an ideal op-amp is zero if the voltage difference between both inputs is zero. This can occur when (1) both inputs are connected to ground, or (2) both inputs are connected to the same voltage source. With a real op-amp however, it is not uncommon to find a small DC output voltage present in either of these two cases. This undesired output voltage level is called the *output offset voltage* (V_{os}), and can result from one or more of the following sources:

1. Input offset voltage.
2. Input bias current.
3. Input offset current.

Although the output offset voltage is usually very small, it can contribute significant errors to the performance of low-level linear amplifier circuits. We will look at all these sources, the calculation of their relative contributions to the total output offset voltage, and practical methods for their elimination.

Input Offset Voltage Component

The input offset voltage (V_{io}) of an op-amp can be represented as a DC voltage source in series with one of its two input terminals. This offset voltage results from the slight mismatches in its internal components. Although it really makes no difference as to which input terminal this equivalent voltage is in series with, it is commonly represented as being in series with the *noninverting input* so that the input offset and resulting output offset have the same polarity.

If the output offset voltage due solely to the input offset voltage is positive with respect to ground, then the DC voltage source representing this input offset voltage has its positive terminal connected to the noninverting input terminal of the op-amp, as shown in Figure 2-6a. On the other hand, if the output offset voltage is negative with respect to ground, then the DC voltage source representing this input offset voltage has its negative terminal connected to the noninverting input terminal of the op-amp, as shown in Figure 2-6b.

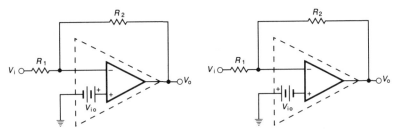

(a) Schematic of inverting amplifier with the positive terminal of the input offset voltage source connected to the noninverting input of the op-amp.

(b) Negative terminal of the input offset voltage source connected to the noninverting input of the op-amp.

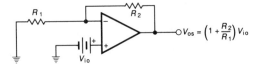

(c) Equivalent circuit having the amplifier input lead grounded.

Fig. 2-6. Contribution of DC input offset voltage.

The equivalent circuit of Figure 2-6c results from either an inverting or a noninverting amplifier having its signal input grounded. For the inverting amplifier, R_1, which is normally connected to the input signal, is now tied to ground. For the noninverting amplifier, the noninverting input, which normally receives the input signal, is now tied to ground. In both cases the equivalent circuit is the same as a nonideal *noninverting* amplifier whose input signal is now the input offset voltage.

Using the closed-loop gain for a noninverting amplifier, the DC output offset voltage of the circuit due solely to this DC input offset voltage component of the nonideal op-amp is

$$V_{os} = A_{CL}V_{io}$$

$$= \left(1 + \frac{R_2}{R_1}\right)V_{io} \qquad \text{(Eq. 2-9)}$$

If a voltage follower is used instead, then the output offset voltage will be the same as the input offset voltage.

For inverting and noninverting amplifiers, the effect of the input offset voltage may be easily minimized by (1) using the lowest possible DC closed-loop gain compatible with system requirements, or (2) using an op-amp with a very low value for V_{io}.

EXAMPLE 2-3

A 741 inverting amplifier has a typical input offset voltage of 2 mV. When used in either an inverting or noninverting amplifier circuit having a closed-loop voltage gain of 15, the typical output offset voltage magnitude would be

$$V_{os} = (1 + 15)(2 \ mV)$$
$$= 32 \ mV \qquad \text{(Eq. 2-9)}$$

The polarity of the resultant output offset voltage is dependent on the polarity of the input offset voltage. For some 741 devices, V_{os} may be positive. For others, it may be negative.

39

Input Bias Current Component

For an ideal op-amp, the input bias currents are zero due to the infinite input impedance. However, input bias currents must be supplied to the two inputs of a real op-amp to provide the proper bias for the differential input stage transistors of the amplifier. These currents, though usually quite small, result in a small output offset voltage. Because of slight mismatches of the input stages of an op-amp, the two bias currents flowing into the two inputs of the op-amp are not equal. Consequently, the input bias offset current is not zero.

The output offset voltage due to the input bias current is then

$$V_{os} = I_B R_2 \qquad \text{(Eq. 2-10)}$$

where R_2 is the feedback resistor of either an inverting amplifier (Fig. 2-3) or noninverting amplifier (Fig. 2-5).

The output offset voltage due to the input bias current can then be minimized by (1) keeping the feedback resistor (R_2) as small as possible for a given closed-loop gain, and/or (2) using an op-amp having low values for input bias current, such as those having FET-input stages.

However, reducing R_2 too much will increase the loading on the op-amp's output and lower the input impedance of the stage. Furthermore, R_2, in the case of an inverting amplifier, may already be fixed at some value by a required input impedance and closed-loop gain. If both the input impedance and closed-loop gain of an inverting amplifier circuit are required to be high, then the output offset voltage due to the input bias current will also be high.

For the inverting amplifier shown in Figure 2-7a, the output offset voltage due solely to the input bias current is eliminated by adding a resistor (R_3) from the amplifier's noninverting terminal to ground. The value of this resistor is chosen so that the current flowing into the noninverting input will exactly offset the current flowing into the inverting input terminal. This required impedance is the parallel combination of R_1 and R_2

$$R_3 = \frac{R_1 R_2}{R_1 + R_2} \qquad \text{(Eq. 2-11)}$$

For a noninverting amplifier, R_3 is placed in series with the noninverting input and the actual signal input (Fig. 2-7b).

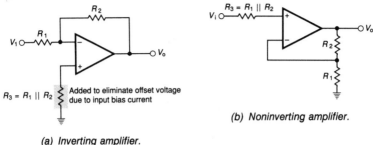

(a) Inverting amplifier.

(b) Noninverting amplifier.

Fig. 2-7. Addition of a resistor to compensate for input bias current error.

Input Offset Current Component

The use of the external resistor to reduce the output offset voltage due to the input bias current is not without its faults. This technique assumed that, at best, the two input bias currents, although not zero, are equal. Therefore, the input offset current would be zero. In virtually all real op-amps, the input offset current is not zero, which in turn produces a DC offset output voltage equal to

$$V'_{os} = I_{os} R_2 \qquad \text{(Eq. 2-12)}$$

where R_2 is the feedback resistor of either an inverting amplifier (Fig. 2-3) or noninverting amplifier (Fig. 2-5).

Since any difference in input bias currents is usually much smaller than the bias currents themselves, the addition of an additional resistor (Fig. 2-7) to reduce the output offset voltage from input bias currents, though not perfect, does reduce the output offset voltage.

EXAMPLE 2-4

The inverting amplifier circuit of Figure 2-8 uses a 741 op-amp having the following typical values:

$$V_{io} = 2 \ mV$$
$$I_B = 80 \ nA$$
$$I_{os} = 2 \ nA$$

The output offset voltage due solely to the input offset voltage is

$$V_{os} = \left(1 + \frac{100\ k\Omega}{10\ k\Omega}\right)(2\ mV)$$
$$= 22\ mV \qquad \text{(Eq. 2-9)}$$

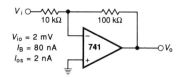

Fig. 2-8. Inverting amplifier circuit example.

The output offset voltage due solely to the input bias current is

$$V_{os} = (80\ nA)(100\ k\Omega)$$
$$= 8\ mV \qquad \text{(Eq. 2-10)}$$

The output offset voltage due solely to the input offset current is

$$V_{os} = (2\ nA)(100\ k\Omega)$$
$$= 0.2\ mV \qquad \text{(Eq. 2-12)}$$

Since it is not known beforehand whether these individual components will be positive or negative voltages, the worst case values must be computed. If all the components were of the same polarity, then the magnitude of the maximum DC output offset voltage possible from all three sources is

$$V_{os}(\max) = 22\ mV + 8\ mV + 0.2\ mV$$
$$= 30.2\ mV$$

On the other hand, some components might be of one polarity, while the remaining would be of the other polarity. Therefore, the magnitude of the minimum DC output offset voltage would be

$$V_{os}(\min) = 22\ mV - (8\ mV + 0.2\ mV)$$
$$= 13.8\ mV$$

so that the possible range of the magnitude of the DC output offset would be 13.8 to 30.2 mV. Using the same op-amp, the output offset could be reduced somewhat by inserting a 9.1-kΩ resistor (using Eq. 2-11) between the noninverting input and ground.

The output offset could be reduced further by reducing the values of R_1 and R_2 to, for example, 1 kΩ and 10 kΩ respectively. The closed-loop voltage gain will still be 10, but the circuit's input impedance will be lowered from 10 kΩ to 1 kΩ. A 910-Ω resistor (if used) would then be inserted between the noninverting input and ground.

OUTPUT OFFSET NULL ADJUSTMENTS

In some op-amp circuits, the output must be DC-coupled to the load. If the DC output offset voltage occurring in the circuit is a critical performance factor, one then has the following options:

1. Select an op-amp with low values of input offset current and input offset voltage.
2. Use minimum resistance values compatible with other circuit requirements.
3. Use the lowest permissible value of closed-loop DC voltage gain.
4. Use a resistor of the proper value in series with the op-amp's noninverting input to cancel the output offset voltage due to the input bias current.

For those cases where these steps do not reduce the output offset sufficiently, some form of external circuitry must then be used to cancel the total DC output offset.

Some op-amp devices, as part of their internal design, have provisions for connecting a *nulling* or *zeroing* multiturn potentiometer to trim any DC output offset voltage to zero when the input signal is zero. The manufacturer's data sheet will show the proper potentiometer value to use and the required connections for a particular type op-amp.

Figure 2-9a illustrates the required connections recommended by manufacturers for the 741 op-amp. Although an inverting amplifier circuit is shown, the placing of a 10-kΩ potentiometer across pins 1 and 5 also applies for noninverting amplifier circuits. By comparison, Figure 2-9b shows the connections for an LM318.

To adjust the potentiometer for zero offset, a digital voltmeter, because of its high resolution (typically ±1 mV), is connected to the circuit's output while grounding the circuit's input. The potentiometer is then carefully adjusted until the voltmeter reads zero.

For those op-amps not having connections for nulling potentiometers, the circuits of Figure 2-10 are used for inverting amplifi-

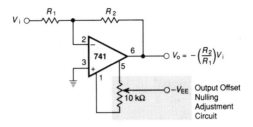

(a) Connections required for a 741 wired as an inverting amplifier.

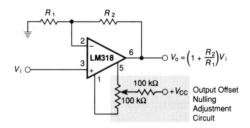

(b) Connections required for an LM318 wired as a noninverting amplifier.

Fig. 2-9. Output offset voltage nulling with op-amps having internal design provisions requiring a single external potentiometer.

ers, noninverting amplifiers, and voltage followers. In all of these circuits, the voltage at the wiper of the potentiometer can vary from $-V_{EE}$ to $+V_{CC}$. When the potentiometer is properly adjusted, resistors R_A and R_B form a voltage divider network that reduces the potentiometer voltage by an amount equal to the DC output voltage, but of the opposite polarity.

For the voltage follower of Figure 2-10d, the voltage developed across R_C cancels the offset. In terms of the three resistances, the output voltage is given by

$$V_o = \left(1 + \frac{R_C}{R_B + (R_A/4)}\right) V_i \qquad \text{(Eq. 2-13)}$$

The same procedure used for adjusting the nulling potentiometer is the same as that for op-amps having internal provisions for zeroing the offset voltage.

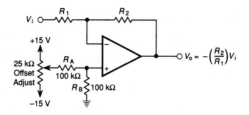

(a) Inverting amplifier.

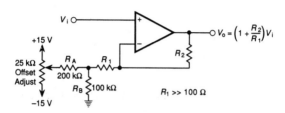

(b) Voltage follower.

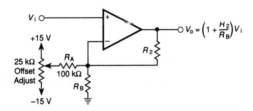

(c) Noninverting amplifier.

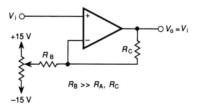

(d) Alternate method for voltage followers.

Fig. 2-10. Output offset voltage nulling with op-amps not having internal design provisions.

45

COMMON-MODE REJECTION RATIO

For the ideal op-amp, the common-mode rejection ratio (CMRR) is infinite, so that any signal that is simultaneously present at both the inverting and noninverting inputs will be eliminated. Figure 2-11 shows a difference amplifier with both inputs tied together. Since $V_1 = V_2$, the input signal, which could be 60-Hz noise, is called the *common-mode input* (V_{icm}). Since the difference amplifier amplifies the voltage difference between its two inputs, the output voltage ideally then is zero. This follows from the ideal input-output voltage equation

$$V_o = \frac{R_2}{R_1}(V_2 - V_1) \qquad \text{(Eq. 2-14)}$$

where the ratio R_2/R_1 is the amplifier's *differential gain* (A_D).

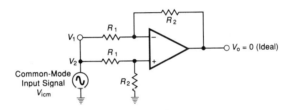

Fig. 2-11. Illustration of common-mode rejection ratio in a difference amplifier with both inputs tied together.

For an actual op-amp like the 741, there will nevertheless be a small *common-mode output* or *error* signal component having the same frequency as the common-mode input signal in addition to no phase shift. The *common-mode gain* (A_{cm}) is the ratio of the common-mode output voltage to the common-mode input voltage

$$A_{cm} = \frac{V_{ocm}}{V_{icm}} \qquad \text{(Eq. 2-15)}$$

The CMRR is based on both the common-mode and differential gains

$$CMRR = \frac{A_D}{A_{cm}} \qquad \text{(Eq. 2-16)}$$

In terms of the op-amp and circuit parameters, the common-mode output voltage is

$$V_{ocm} = \left(\frac{R_2}{R_1 CMRR}\right) V_{icm} \qquad \text{(Eq. 2-17)}$$

and is solely due to a common-mode input signal. The total output voltage of a nonideal difference amplifier is the sum of the output of the ideal amplifier (Eq. 2-14) and the common-mode output component (Eq. 2-17)

$$V_o = \frac{R_2}{R_1}(V_2 - V_1) + \left(\frac{R_2}{R_1 CMRR}\right) V_{icm} \qquad \text{(Eq. 2-18)}$$

As increasing the CMRR will reduce the common-mode error voltage component, op-amps used in difference amplifier circuits should then have the highest CMRR possible.

In many cases however, simply selecting an op-amp with a high CMRR may not sufficiently reduce the common-mode error component to a level where it does not severely affect the amplifier's performance. This is particularly true if the differential input signal levels are very small in comparison with a common-mode input signal level. Trying to zero-out the common-mode output with any of the offset methods discussed earlier in this chapter will not work. Remember, output voltage offset is a *DC* phenomenon while the common-mode error is primarily an *AC* problem.

One effective method of increasing the difference amplifier's common-mode rejection is the addition of a potentiometer as shown in Figure 2-12a. Even though the resistance ratio R_2/R_1 for both the inverting and noninverting inputs may be exactly equal, slight tolerances in manufacturing still cause an imbalance in the input currents of the op-amp. If both inputs are tied together to an AC source, such as 60 Hz as shown in Figure 2-12b, the potentiometer is then adjusted so that the resultant common-mode output signal is zero, or a minimum.

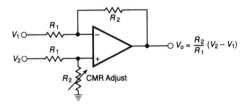

(a) Using a potentiometer.

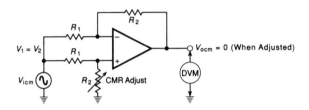

(b) Adjustment of potentiometer for minimum output voltage with both inputs tied together to a common-mode input voltage.

Fig. 2-12. *Increasing the common-mode rejection of an op-amp.*

EXAMPLE 2-5

The difference amplifier of Figure 2-13 is used to determine the common-mode rejection of a given 741 op-amp, using the following measured values:

$$V_{icm} = 5.73 \text{ V RMS}$$
$$V_{ocm} = 0.29 \text{ V RMS}$$

The common-mode gain is

$$A_{cm} = \frac{0.29 \text{ V}}{5.73 \text{ V}} \quad \text{(Eq. 2-15)}$$
$$= 0.051$$

The differential gain of the difference amplifier circuit is

$$A_D = \frac{100 \text{ k}\Omega}{100 \Omega}$$
$$= 1000$$

so that the CMRR is

$$CMRR = \frac{1000}{0.051}$$
$$= 19{,}608$$
(Eq. 2-16)

When expressed in decibels, the common-mode rejection (CMR) is

$$CMR_{dB} = 20 \log_{10}(19{,}608)$$
$$= 85.8 \; dB$$
(Eq. 2-2)

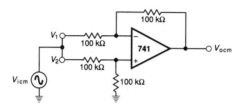

Fig. 2-13. Difference amplifier circuit example.

THE EFFECT OF FREQUENCY ON CIRCUIT PERFORMANCE

One characteristic of the ideal op-amp discussed in Chapter 1 is that its bandwidth is infinite. That is, it has no upper frequency limit. Unfortunately, the bandwidth of real op-amps is limited and there are several concepts that must be kept in mind to insure that the particular op-amp circuit works properly as intended at all frequencies of interest.

Open-Loop Gain and Gain-Bandwidth Product

Figure 2-14 shows the frequency response of the open-loop voltage gain of a typical op-amp. At DC and frequencies below approximately 10 Hz, A_{OL} is constant at +100 dB, equivalent to an ordinary voltage gain of 100,000. Above a cutoff frequency of 10 Hz, the response drops off at a linear rate of −20 dB/decade, or −6 dB/octave. This decrease continues until the open-loop gain is unity. The frequency at which this occurs is the *unity-gain frequency* (f_T), which is also called the *unity-gain crossover frequency* as well as the *small-signal unity-gain bandwidth*.

Since the open-loop gain varies with frequency, those parameters that are dependent on the open-loop gain are also dependent on

49

frequency. Those parameters discussed so far include the amplifier's closed-loop input and output impedances, and to a certain extent, the common-mode rejection. In most cases, it is not necessary to have a plot of the open-loop gain response. Instead, a more useful parameter is the *gain-bandwidth product* (GBP), which is a constant for a given op-amp. For the 741, it is 1 MHz, while the LM318 has a GBP of 15 MHz. It is dependent on the op-amp's open-loop gain and the required bandwidth

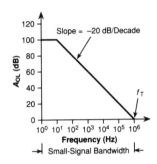

Fig. 2-14. Open-loop frequency response of a typical op-amp.

$$GBP = A_{OL}BW \qquad \text{(Eq. 2-19)}$$

Using Equation 2-19, the open-loop gain for frequencies above its cutoff frequency can be calculated for a given bandwidth. Since the open-loop and closed-loop gains are interrelated, the GBP also specifies the maximum closed-loop gain possible.

EXAMPLE 2-6

The LM318 op-amp has a GBP of 15 MHz. If the expected maximum input frequency is 30 kHz, then the LM318's open-loop gain is

$$A_{OL} = \frac{15\ MHz}{30\ kHz}$$
$$= 500\ (or\ +54\ dB)$$

As shown in Figure 2-15, the op-amp's bandwidth is the point at which the closed-loop voltage gain curve intersects the open-loop gain curve. Here are plotted a family of closed-loop voltage gain curves for a 741 op-amp. For a closed-loop gain of 100 (+40 dB), the maximum bandwidth possible is then 10 kHz. However, if the circuit's closed-loop gain is reduced to 10 (+20 dB), then the maximum bandwidth possible is increased to 100 kHz. Consequently, the tradeoff of lower gain allows for a larger bandwidth, and vice versa.

To be consistent with good engineering practices, it is never a good idea to operate any circuit or device at its limits. Conse-

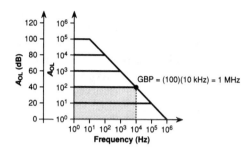

Fig. 2-15. Illustration of the gain-bandwidth product of a 741 op-amp.

quently, the maximum closed-loop gain at the most should be limited to about one-half the available open-loop gain to give a reasonable degree of safety margin. If the GBP of a given op-amp is 15 MHz, the maximum open-loop gain at 30 kHz is then 15 MHz/30 kHz, or 500. The maximum allowable closed-loop gain permissible problem free performance should then be ½ × 500, or 250. Of course, the closed-loop gain can be less than 250, but it should not exceed 250.

Rise Time

Some data sheets may not give either the open-loop gain response curve, the gain-bandwidth product, or the unity-gain frequency. However, a parameter called the *unity-gain transient response rise time* (t_r) is given. This is related to the op-amp's unity-gain bandwidth (or GBP) by

$$GBP = \frac{0.35}{t_r} \qquad \text{(Eq. 2-20)}$$

For the 741, t_r is typically 0.3 μs, giving a GBP of 0.35/0.3 μs or 1.16 MHz. Although there may be a slight difference between calculating the GBP using the rise time and the value given in data sheet, the two can be considered equal for all practical purposes.

Slew Rate Limiting

In addition to the open-loop gain response curve and GBP, the op-amp's slew rate is another factor limiting the frequency response. For high-frequency, large output voltage swings, the slew rate is a greater factor than the gain-bandwidth product.

The maximum time rate of change that the output signal is capable of is dependent on both its peak amplitude and frequency. For a sine wave, the maximum rate of change occurs at the zero crossing. If this rate of change is greater than the slew rate of the op-amp, then the output signal will be distorted somewhat. For a sine-wave input, the distorted output appears like a triangle waveform (Fig. 2-16). Here in effect, the slew rate limits how fast the output sine-wave signal can change with time.

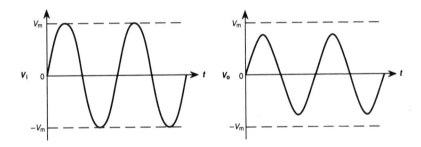

Fig. 2-16. Slew rate distortion of a sine wave.

For a given slew rate (SR) and peak output voltage (V_P), the maximum input sine-wave frequency without causing the output signal to take on a triangular shape is given by

$$f_{max} = \frac{SR}{2\pi V_P} \qquad \text{(Eq. 2-21)}$$

Many manufacturers, as a matter of convenience, often use either the term *full-power bandwidth* or *large-signal response* in their data sheets to represent f_{max}.

EXAMPLE 2-7

The 741 has a slew rate of 0.5 V/μs, and when using a ±9-V supply, the maximum output voltage swing capable is about ±7 V. The full-power bandwidth for the 741 is

$$f_{max} = \frac{0.5\ V/\mu s}{2\pi(7\ V)}$$ (Eq. 2-21)
$$= 11.4\ kHz$$

FREQUENCY COMPENSATION

Despite some of their nonideal characteristics which present some minor drawbacks, general-purpose op-amps like the 741 nevertheless are very easy to work with. As long as we keep the gain-bandwidth product rule in mind, not much will go wrong.

The 741 is an example of an *internally compensated* op-amp. This means that the manufacturer has already optimized its design to give a controlled roll-off characteristic to prevent the device from oscillating. Inside the 741 is a *compensating capacitor* which controls the frequency stability, and obviously cannot be removed or disconnected. In exchange for this hassle-free device, one settles for the following tradeoffs:

1. Reduced gain-bandwidth product.
2. Slower slew rate.
3. Reduced full-power bandwidth.
4. Higher rise time.

To achieve a higher performance level, one either has to use a compensated op-amp with higher ratings, or use an uncompensated op-amp and provide some amount of external compensation. No op-amp can be left without some form of compensation since there will be a very strong likelihood that high-gain, high-frequency amplifier circuits would become unstable and self-oscillate. In an uncompensated device, the terminals for the frequency compensating capacitor are brought out to allow the user to select the optimum parameters of bandwidth and stability by connecting one or more external capacitors.

As a practical matter, the data sheets of uncompensated op-amps give the details as to how to determine the amount of compensation for a required gain and frequency response.

The methods of compensation vary from one type op-amp to another. For example, the LM301 is a typical uncompensated, general-purpose op-amp. As shown in Figure 2-17a, one form of external compensation is achieved by connecting a single capacitor (C_1) between pins 1 and 8. If $C_1 = 3$ pF, the bandwidth is approximately 10 MHz. Increasing the value of C_1 in turn decreases the bandwidth, as well as full-power bandwidth. If $C_1 = 30$ pF, then the bandwidth decreases to approximately 1 MHz, similar to the internally compensated 741. Figure 2-17b graphs the effect of the compensating capacitor on the op-amp's bandwidth.

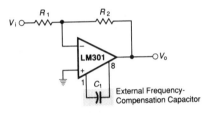

(a) External connections.

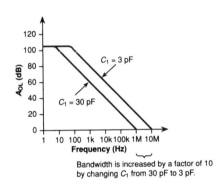

Bandwidth is increased by a factor of 10 by changing C_1 from 30 pF to 3 pF.

(b) Open-loop frequency response.

Fig. 2-17. Frequency compensation of an LM301 op-amp.

Chapter 3

Differentiators and Integrators

INTRODUCTION AND OBJECTIVES

The operational amplifier is capable of performing two additional mathematical operations of *differentiation* and *integration*. These two functions are the mathematical inverses of each other, just as multiplication is the inverse of division. Although the electronic circuits that perform these operations (*differentiators* and *integrators* respectively) are not very complex, they proved indispensable in solving calculus-based equations during the days of the analog computer.

Besides increasing the computational power of the analog computer, differentiators and integrators serve other useful purposes. Very often these circuits are used to further process certain signals, such as a detector in some FM modulators changing a triangle wave into a square wave, a process often refered to as *waveshaping*. Elsewhere, integrators are a fundamental building block of several types of waveform generators (Chapter 8), as well as several active filters (Chapter 9).

At the completion of this chapter, you will be able to:

- Determine the output voltage of the basic op-amp differentiator.
- Explain why it is often necessary to stabilize the basic differentiator circuit and how this is accomplished.
- Determine the output voltage of the basic op-amp integrator.
- Explain why it is often necessary to stabilize the basic integrator circuit and how this is accomplished.

- For several common input waveforms, show what the output waveforms from both a differentiator and integrator would look like.

THE DIFFERENTIATOR

The *differentiator*, not to be confused with a differential or difference amplifier, is a circuit whose output voltage is proportional to the instantaneous time rate of change* of its input signal

$$V_o(t) = k\left(\frac{\Delta V_i}{\Delta t}\right) \quad \text{(Eq. 3-1)}$$

where k is a constant of proportionality. The operation of a differentiator is the same as taking the "slope of the line" at any point on the input signal. For input signals not having straight line segments, such as a sine wave, the slope at any point on that curve is the slope of the line that is tangent to that point.

If the amplitude of an input signal is not changing with time, such as a DC voltage, then its slope is zero and the differentiator produces no output signal. For the most part, it is not the amplitude of the input signal that is important, but rather *the time rate of change*. Differentiators find many uses in electronic circuits, such as those which measure the rate of change of a signal, waveshaping, and producing control signals that coincide with rapid changes in signal level.

The Basic Circuit

Figure 3-1 shows the basic operational amplifier differentiator. It is similar in form to the inverting amplifier except that the input element is a *capacitor* instead of a resistor. The output voltage is given by

$$V_o = -R_F C\left(\frac{\Delta V_i}{\Delta t}\right) \quad \text{(Eq. 3-2)}$$

The quantity $\Delta V_i/\Delta t$ represents the time rate of change, or slope of the input signal over any small change in time. In the notation of differential calculus, Equation 3-2 is written as

---

* In calculus, this operation is the taking the first derivative of the input voltage with respect to time, dV_i/dt.

$$V_o = -R_F C \left(\frac{dV_i}{dt}\right) \qquad \text{(Eq. 3-3)}$$

The quantity $R_F C$ is simply a scaling factor, and equals the proportionality constant k in Equation 3-1. This quantity is often referred to as the *differentiator gain*. The minus sign in Equation 3-3 implies that the polarity of the differentiated output signal is the opposite of what it normally would be. If the input signal is increasing with time (i.e., a positive slope), the resulting differentiated output signal will be negative. If the input signal is decreasing with time, the resulting output signal will be positive.

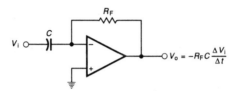

Fig. 3-1. Basic differentiator.

EXAMPLE 3-1

A 1-kHz triangle wave having a peak amplitude of 1 V is the input signal to the differentiator circuit of Figure 3-2a.

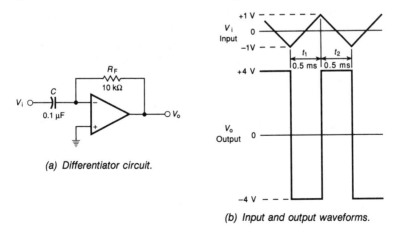

Fig. 3-2. Example of slope calculation.

The input waveform must be analyzed in two sections covering time periods t_1 and t_2. Since the period of one cycle is 1/1 kHz, or 1 ms and the triangle wave is symmetrical, both t_1 and t_2 are 0.5 ms. For t_1, the positive-going slope is

$$slope\,(t_1) = \frac{2V}{0.5\ ms}$$
$$= 4000\ V/s$$

In similar fashion, the slope of the negative-going portion during t_2 is

$$slope\,(t_2) = \frac{-2V}{0.5\ ms}$$
$$= -4000\ V/s$$

Using Equation 3-2, the output voltage of the differentiator for time period t_1 is

$$V_o(1) = -(10\ k\Omega)(0.1\ \mu F)(4000\ V/s)$$
$$= -4\ V$$

while for time period t_2

$$V_o(2) = -(10\ k\Omega)(0.1\ \mu F)(-4000\ V/s)$$
$$= +4\ V$$

Figure 3-2b shows the corresponding input and output waveforms for the differentiator circuit. Note that the positive-going slope is constant for period t_1 and the output signal equals a constant negative voltage. For period t_2, the negative-going constant slope produces a constant positive voltage of +4 V.

The basic differentiator works properly for all input signals with frequencies greater than

$$f_1 = \frac{1}{2\pi R_F C} \qquad \text{(Eq. 3-4)}$$

The Frequency Compensated Differentiator

One serious problem with the basic differentiator circuit is that it is very susceptible to high-frequency electrical noise. This is because the reactance of the capacitor decreases with frequency causing a corresponding increase in the closed-loop voltage gain at the rate of +6 dB/octave. Although the basic differentiator's closed-loop gain increases with frequency, it is limited at the high end by the op-amp's open-loop response curve. For frequencies above where the response intersects the open-loop curve, the circuit stops acting as a differentiator.

To put a limit on the closed-loop gain at high frequencies before being limited by the op-amp's open-loop response curve, resistor R_S is added in series with the input capacitor as shown in the stabilized, compensated circuit of Figure 3-3a. This type of compensated differentiator circuit has a much improved noise handling ability, but the maximum usable input frequency over which the improved circuit acts as a differentiator is now limited to input frequencies below

$$f_2 = \frac{1}{2\pi R_S C} \qquad \text{(Eq. 3-5)}$$

This frequency must be greater than the frequency where the differentiator's low frequency closed-loop gain is unity (0 dB), such that

$$f_2 > \frac{1}{2\pi R_F C} \qquad \text{(Eq. 3-6a)}$$

or

$$\frac{1}{2\pi R_S C} > \frac{1}{2\pi R_F C} \qquad \text{(Eq. 3-6b)}$$

At frequencies above f_2, the capacitor's reactance is very small compared with R_S so that the circuit essentially looks like an inverting amplifier (Fig. 3-3b) having a closed-loop voltage gain equal to

$$A_{CL} = -\frac{R_F}{R_S} \qquad \text{(Eq. 3-7)}$$

Figure 3-3c graphs resultant frequency response of the frequency-compensated differentiator in relation to the op-amp's open-loop voltage gain.

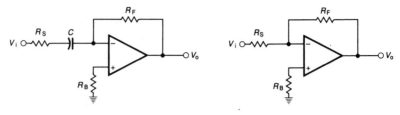

(a) Basic circuit.   (b) Equivalent high-frequency circuit.

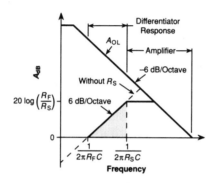

(c) Frequency response graph showing high-frequency limiting.

Fig. 3 3. Compensated differentiator

From the gain-bandwidth product, the frequency at which closed-loop and open-loop gain curves intersect equals the op-amp's unity-gain frequency (f_T) divided by the closed-loop gain. For input frequencies between f_1 and f_2, the improved circuit works properly as a differentiator. Frequency f_2 is set approximately one decade above highest frequency that the circuit is to properly differentiate input signals.

As an example, if the highest expected input signal frequency to be differentiated is 500 Hz, then f_2 should be set to approximately 10 × 500 Hz, or 5 kHz for proper operation. As a general procedure, the design of any circuit having capacitors is best accomplished by first selecting a given standard value for the capacitor(s) and then calculating the resulting value(s) for the resistors. This is done because the range of standard capacitor values is not as extensive as it is for those resistors having 5 or 10 percent tolerances.

As with many linear op-amp circuits, resistor R_B is often used to reduce the effects of the op-amp's input bias current. Since the capacitor blocks DC current, R_B in practice is made equal to R_F. In many situations however, R_B can be omitted with the noninverting input connected directly to ground.

For use as a differentiator, the op-amp should have a high slew rate in order to properly respond to very quickly changing signals, as well as a high gain-bandwidth product. The use of high-quality capacitors, such as polystyrene and Teflon, are preferred.

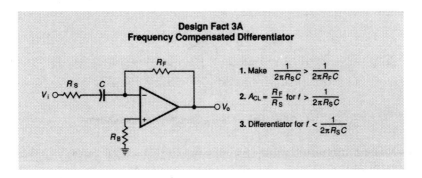

Design Fact 3A
Frequency Compensated Differentiator

1. Make $\dfrac{1}{2\pi R_S C} > \dfrac{1}{2\pi R_F C}$

2. $A_{CL} = \dfrac{R_F}{R_S}$ for $f > \dfrac{1}{2\pi R_S C}$

3. Differentiator for $f < \dfrac{1}{2\pi R_S C}$

EXAMPLE 3-2

Determine the necessary component values required for the frequency-compensated differentiator circuit of Figure 3-3a to properly differentiate input signals up to 2 kHz. Assume that the op-amp's gain-bandwidth product is 1 MHz.

For proper differentiation of signals up to 2 kHz, f_1 should be set approximately one decade higher, or 20 kHz. Choosing a standard value for C, such as 0.033 µF, then

$$R_S = \dfrac{1}{2\pi(20\ kHz)(0.033\ \mu F)}$$ (Eq. 3-5)

= 241 Ω (*use* 240 Ω *standard value*)

Above 20 kHz, the closed-loop voltage gain should be limited to some convenient value consistent with the gain-bandwidth product of the op-amp used. The closed-loop voltage gain allowable is then

$$A_{CL} = \dfrac{1\ MHz}{20\ kHz}$$ (Eq. 2-19)

= 50

61

The value for the high-frequency closed-loop gain should be less than this maximum value. For example, choosing a gain of 50, then

$$R_F = (50)(240\Omega)$$
$$= 12 \ k\Omega \quad (a \ standard \ value) \quad \text{(Eq. 3-7)}$$

The final circuit is shown in Figure 3-4.

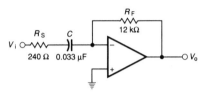

Fig. 3-4. Completed example circuit for a differentiator.

The minimum frequency above which this circuit works as a differentiator is

$$f_1 = \frac{1}{2\pi(12 \ k\Omega)(0.033 \ \mu F)} \quad \text{(Eq. 3-4)}$$
$$= 402 \ Hz$$

and behaves as an inverting amplifier for input frequencies greater than 20 kHz.

INTEGRATORS

The mathematical inverse of differentiation is integration, and the circuit that performs this is called an *integrator*. Integration is the mathematical process of determining the accumulative area underneath the curve or upper boundary of waveform over a given period of time. The electronic integrator that is built around an op-amp produces an output voltage signal proportional to the area underneath the boundary formed by the input voltage signal.

The Basic Circuit

Owing to the inverse relationship between integration and differentiation, we create the basic op-amp integrator of Figure 3-5 by simply interchanging the positions of the resistor and capacitor of the basic differentiator circuit (Fig. 3-1). Here, the resistor is now the input element while the capacitor becomes the feedback element.

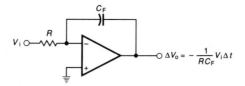

Fig. 3-5. Basic integrator.

The change in the output voltage is given by

$$\Delta V_o = -\left(\frac{1}{RC_F}\right) V_i \Delta t \quad \text{(Eq. 3-8)}$$

The $1/RC_F$ term in Equation 3-8 is a scaling factor, called the *integrator gain*. The minus sign tells us that the polarity of the output voltage representing the integral, or accumulated area underneath the curve of the input signal over one cycle, is of the opposite polarity from the normal sense.

In the notation of integral calculus, the output voltage of the integrator of a time dependent input voltage V_i over the time period from t_1 to t_2 is written as

$$V_o = \int_{t_1}^{t_2} V_i \, dt \quad \text{(Eq. 3-9)}$$

The basic integrator works properly for all input signals with frequencies less than

$$f_2 = \frac{1}{2\pi RC_F} \quad \text{(Eq. 3-10)}$$

The Frequency Compensated Integrator

Like the differentiator, the basic integrator circuit presents its own problems under certain conditions. First, any of the nonideal characteristics of the op-amp that produces an output offset voltage with no applied input signal will also be integrated over the same period of time. If left unchecked, the output voltage will either slowly rise from zero towards the positive supply voltage or decrease towards the negative supply voltage. In either case, this action eventually saturates the op-amp. Secondly, the reactance of the feedback capacitor varies with frequency. At very low frequencies, the closed-loop gain becomes very large and approaches the op-amp's open-loop gain.

A stabilized, compensated integrator circuit that overcomes or at least minimizes these problems is shown in Figure 3-6a. To limit the DC closed-loop gain to a reasonable value (typically 10 to 100), resistor R_F is placed in parallel with the feedback capacitor. At DC, the equivalent circuit, shown in Figure 3-6b looks like an inverting amplifier with a closed-loop voltage gain of

$$A_{CL} = -\frac{R_F}{R} \qquad \text{(Eq. 3-11)}$$

In addition to limiting the closed-loop gain, R_F also reduces the output offset voltage due to the op-amp's input offset voltage. However, there is a price that is paid for these improvements. Resistor R_F now limits the use of the integrator to those input frequencies above

$$f_1 = \frac{1}{2\pi R_F C_F} \qquad \text{(Eq. 3-12)}$$

This frequency must be less than the frequency where the differentiator's low-frequency closed-loop gain is unity (f_2), such that

$$f_1 < \frac{1}{2\pi RC_F} \qquad \text{(Eq. 3-13a)}$$

or

$$\frac{1}{2\pi R_F C_F} < \frac{1}{2\pi RC_F} \qquad \text{(Eq. 3-13b)}$$

As a practical matter, f_1 should be set approximately one decade below the lowest anticipated input frequency. For input frequencies below f_1, the circuit looks like an inverting amplifier with a closed-loop voltage gain equal to $-R_F/R$ (Fig. 3-6c). For input frequencies from f_1 to f_2, the compensated circuit works properly as an integrator. Furthermore, the input impedance of the integrator is set by the value of R.

To reduce the effects of the input bias current from possibly saturating the op-amp, resistor R_B is connected between the op-amp's

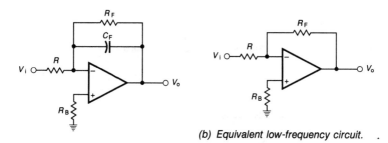

(a) Basic circuit.

(b) Equivalent low-frequency circuit.

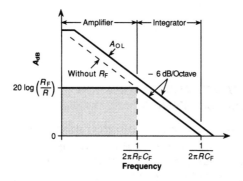

(c) Frequency response graph showing low-frequency limiting.

Fig. 3-6. Compensated integrator.

noninverting input and ground. The value for R_B is equal to the parallel combination of R_F and R

$$R_B = \frac{R_F R}{R_F + R} \quad \text{(Eq. 3-14)}$$

For best overall performance, the feedback capacitor should be a high-quality type having very low leakage such as polystyrene, Teflon, or mica. For relatively short integration times, Mylar capacitors are satisfactory.

The choice of the op-amp will affect the accuracy of operation. Chopper-stabilized op-amps, which serve to reduce DC drift, are preferred for long-term integrators. FET-input devices are used for medium-term periods because of their low input bias current. The standard bipolar-type device is used for very short integration times, such as those in certain active filters and most audio applications.

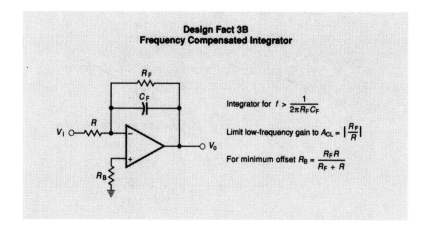

DIFFERENTIATION AND INTEGRATION OF COMMON WAVEFORMS

Figure 3-7 shows the effects of integration and differentiation on some common waveforms. Differentiating a sine wave results in another sinusoidal waveform advanced in time by 90°, or a cosine wave. Integration of a sine wave produces another sinusoidal waveform retarded in time by 90°, again a cosine. Since the circuit for either operation causes a polarity inversion, both waveforms show a 180° phase shift.

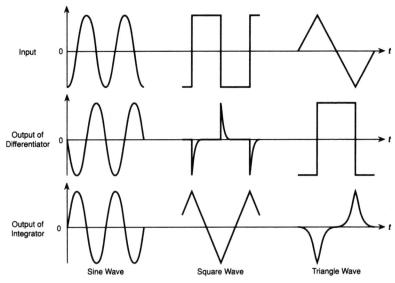

Fig. 3-7. Differentiation and integration of common waveforms.

Differentiating a square wave causes a series of spikes to be generated. Because of the polarity inversion of the circuit, differentiation of the positive cycle of the square results in a negative-going spike, and vice versa. A triangle wave can be obtained by integrating a square wave. The triangle output has a positive slope when the square wave voltage is negative. The triangle output has a negative slope during the time the input square wave voltage is positive.

Differentiating a triangle wave gives a square wave. As the triangle wave slopes negative, the square wave output is positive. As the triangle wave slopes positive, the square wave output is negative. Integrating a triangle wave yields a distorted triangle wave having curved sides. Note that this output is 180° out of phase with the input signal.

Chapter 4

Single Supply Operation

INTRODUCTION AND OBJECTIVES

Although primarily designed to be used with a bipolar, or split power supply, many op-amp circuits can be powered by a single supply voltage. The cost of implementing a given op-amp circuit using two power supplies is higher than one that could be built using a single-ended power supply. For battery-powered portable circuits, the extra battery required for conventional split supply operation not only adds to the cost of the circuit, it also adds some weight.

This chapter discusses how many of the bipolar linear op-amp circuits discussed in earlier chapters can be externally biased to operate using a single supply voltage. Because these circuits use capacitor coupling to remove unwanted DC levels, single supply biased circuits are designed only for AC operation.

At the completion of this chapter, you will be able to design and predict the operation of a bipolar op-amp using a single supply voltage in the following AC-coupled circuits:

- voltage follower
- inverting and noninverting amplifiers
- summing amplifier
- difference amplifier

SINGLE SUPPLY BIASING

The op-amp, using a single supply voltage, must be able to produce both negative-going and positive-going signals. The best approach is to set the op-amp's DC output voltage equal to one-half the positive supply voltage, $V_{CC}/2$ with no input signal applied. When an input signal (such as a sine wave) is applied to a single supply biased circuit, the output voltage will then vary about the DC voltage level of $V_{CC}/2$ when no signal is applied (Fig. 4-1). The resultant sine-wave signal is now a superposition of both the DC voltage component and the amplified AC signal. For proper amplifier operation, this DC component must then be removed from the output signal before the desired AC output is coupled to a load.

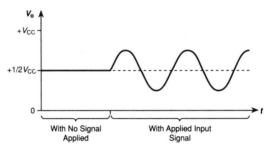

Fig. 4-1. The quiescent DC output biased to one-half the supply voltage.

VOLTAGE FOLLOWER

The basic bias, or DC circuit for the voltage follower powered by a single supply voltage is shown in Figure 4-2a. Resistors R_1 and R_2, whose values are made equal, form a voltage divider whose DC voltage across R_2 is one-half the supply voltage V_{CC}. The op-amp's negative supply pin is grounded and, contrary to the convention for split supplies, the power supply connections for single supply biasing are not omitted from schematics.

Since the closed-loop voltage gain for the voltage follower is 1, and the DC input voltage at the op-amp's noninverting input is one-half the supply voltage, the DC output voltage is also one-half the supply voltage.

Figure 4-2b shows the completed circuit for the single supply biased AC voltage follower. As in the basic bias circuit, R_1 and R_2 form the voltage divider that puts the op-amp's noninverting input at one-half the supply voltage. Resistors R_1 and R_2 can be any value as long as they are equal. In practice, they are chosen to be in the 10-kΩ to 30-kΩ range.

Resistor R_3 is not really necessary as it only conveniently sets the input impedance of the circuit. When used, it is made at least ten times larger than R_1 and R_2. If it is omitted along with C_2 as not to

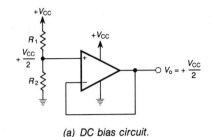

(a) DC bias circuit.

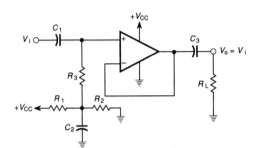

(b) Complete circuit.

Fig. 4-2. Single supply voltage follower.

short the AC signal to ground, the voltage follower's input impedance is then essentially equal to R_1 in parallel with R_2. Capacitor C_2 is a *bypass*, or *decoupling capacitor*, acting as a low-pass filter to remove interfering low-frequency signals (such as 60 Hz) from reaching the op-amp's input via the power supply.

The cutoff frequency (f_2) above which this RC network sets the junction of R_1, R_2, R_3, and C_2 at AC ground is

$$f_2 = \frac{R_1 + R_2}{2\pi R_1 R_2 C_2} \qquad \text{(Eq. 4-1)}$$

The cutoff frequency (f_1) for the R_3-C_1 input network is found from

$$f_1 = \frac{1}{2\pi R_3 C_1} \qquad \text{(Eq. 4-2)}$$

Any AC input signal above this frequency will then be superimposed on the DC bias voltage, producing an identical output signal that is in phase with the AC input signal.

The cutoff frequency (f_3) of the output network is determined by C_3 and R_L, which is either the load or the input impedance of the next stage

$$f_3 = \frac{1}{2\pi R_L C_3} \qquad \text{(Eq. 4-3)}$$

As a general rule for simplicity, all three cutoff frequencies are made approximately equal and are set at approximately one-tenth the lowest input frequency to be amplified.

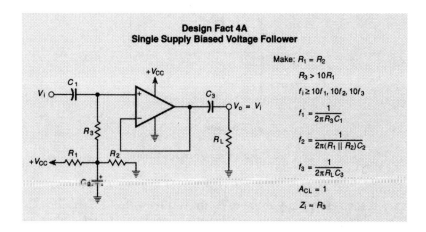

Design Fact 4A
Single Supply Biased Voltage Follower

Make: $R_1 = R_2$
$R_3 > 10 R_1$
$f_i \geq 10 f_1, 10 f_2, 10 f_3$
$f_1 = \frac{1}{2\pi R_3 C_1}$
$f_2 = \frac{1}{2\pi (R_1 \| R_2) C_2}$
$f_3 = \frac{1}{2\pi R_L C_3}$
$A_{CL} = 1$
$Z_i \approx R_3$

EXAMPLE 4-1

Using the voltage follower circuit of Figure 4-3, determine the following:

- The quiescent DC output voltage.
- The circuit's input impedance.
- The low-frequency response.

Based on the two 10-kΩ resistors and the +15-V supply, the DC voltage at the op-amp's noninverting input is one-half the supply voltage, or +7.5 V. Since the voltage follower's closed-loop voltage gain is 1, the DC quiescent output voltage is also +7.5 V, while the input impedance of the voltage follower circuit is equal to the 180-kΩ resistor.

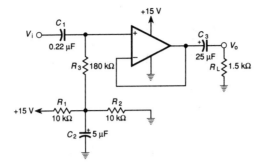

Fig. 4-3. Single supply biased voltage follower circuit.

The cutoff frequencies are

$$f_1 = \frac{1}{2\pi(180\ k\Omega)(0.22\ \mu F)}$$
$$= 4\ Hz$$

(Eq. 4-2)

$$f_2 = \frac{(10\ k\Omega + 10\ k\Omega)}{2\pi(10\ k\Omega)(10\ k\Omega)(5\ \mu F)}$$
$$= 6.4\ Hz$$

(Eq. 4-1)

and

$$f_3 = \frac{1}{2\pi(1.5\ k\Omega)(25\ \mu F)}$$
$$= 4.3\ Hz$$

(Eq. 4-3)

Of these three cutoff frequencies, the one associated with the voltage divider network ($f_2 = 6.4$ Hz) is the *dominant* (i.e., highest) one, so that the circuit will operate properly as a single supply biased voltage follower for AC input signals approximately 10 times above this frequency, or 64 Hz.

NONINVERTING AMPLIFIER

Figure 4-4 shows the circuit for a noninverting amplifier biased with a single supply voltage. The technique for biasing this circuit is identical to that used for the single supply biased voltage follower. Resistors R_1 and R_2 form the required voltage divider which puts

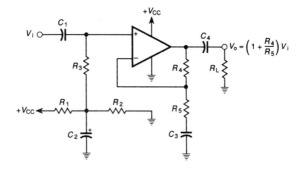

Fig. 4-4. Single supply biased noninverting amplifier.

the noninverting input at a DC level equal to one-half the power supply voltage while C_2 provides power supply noise decoupling. The input impedance is equal to R_3 if both R_1 and R_2 are made at least ten times smaller than R_3.

The cutoff frequency (f_2) for the R_1-R_2-C_2 network is the same as Equation 4-1

$$f_2 = \frac{R_1 + R_2}{2\pi R_1 R_2 C_2} \qquad \text{(Eq. 4-4)}$$

The cutoff frequency (f_1) for the R_3-C_1 input network is

$$f_1 = \frac{1}{2\pi R_3 C_1} \qquad \text{(Eq. 4-5)}$$

For AC input signals, C_3 effectively grounds the lower end of R_5 so that the closed-loop voltage gain for AC input signals is the same as for noninverting amplifiers powered by a split supply

$$A_{CL} = 1 + \frac{R_4}{R_5} \qquad \text{(Eq. 4-6)}$$

and the output signal will be in phase with the input.

The cutoff frequency (f_3) for the R_5-C_3 network is

$$f_3 = \frac{1}{2\pi R_5 C_3} \qquad \text{(Eq. 4-7)}$$

As a general rule, this frequency is made about five to ten times larger than the cutoff frequency associated with the input RC network (f_1) in order to minimize the op-amp's low-frequency gain.
The cutoff frequency for the R_L-C_4 output network is determined in a manner similar to the voltage follower circuit

$$f_4 = \frac{1}{2\pi R_L C_4} \qquad \text{(Eq. 4-8)}$$

and is made the same as f_1 and f_2.

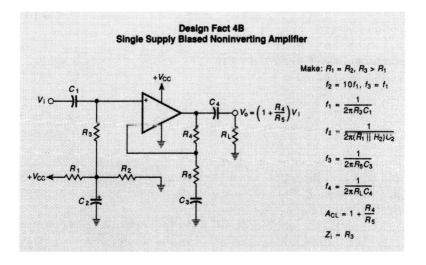

INVERTING AMPLIFIER

Figure 4-5 shows the circuit of an AC inverting amplifier that uses a single-ended power supply. Resistors R_3 and R_4 (10 kΩ to 30 kΩ) are made equal to each other and act as the required voltage divider. The cutoff frequency (f_2) for the R_3-R_4-C_2 bypass network is then

$$f_2 = \frac{1}{\pi R_3 C_2} \qquad \text{(Eq. 4-9)}$$

75

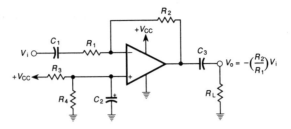

Fig. 4-5. Single supply biased inverting amplifier.

The cutoff frequency (f_1) for the R_1-C_1 input network is

$$f_1 = \frac{1}{2\pi R_1 C_1} \qquad \text{(Eq. 4-10)}$$

For AC input signals, the inverting amplifier has a closed-loop voltage gain equal to

$$A_{CL} = -\frac{R_2}{R_1} \qquad \text{(Eq. 4-11)}$$

As this is an inverting amplifier, the output signal will be 180° out of phase with the input, which is reflected by the use of the minus sign in Equation 4-11. The circuit's AC input impedance is effectively equal to R_1.

The cutoff frequency for the R_L-C_3 output network is determined using Equation 4-3. In general, all three cutoff frequencies are made the same.

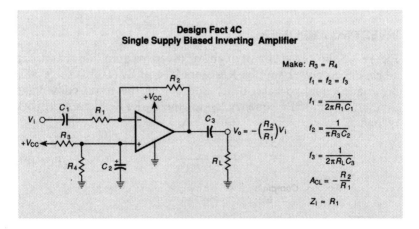

76

EXAMPLE 4-2

The inverting amplifier circuit of Figure 4-4 is to amplify all signals above 50 Hz when connected to a 500-Ω load. Determine the necessary component values to have a closed-loop voltage gain of 20 with an input impedance of 1 kΩ.

We choose a convenient standard value for R_3 and R_4, such as 15 kΩ. As the circuit's input impedance is required to be 1 kΩ, then $R_1 = 1$ kΩ. Also for a closed-loop voltage gain of 20

$$R_2 = (20)(1 \ k\Omega)$$
$$= 20 \ k\Omega \quad \text{(Eq. 4-11)}$$

As the minimum expected input frequency is 50 Hz, then each RC network should have a cutoff frequency that is one-tenth this value, or 5 Hz. The capacitor value for each RC network is calculated

$$C_1 = \frac{1}{2\pi(1 \ k\Omega)(5 \ Hz)} \quad \text{(Eq. 4-10)}$$
$$= 32 \ \mu F \quad (use \ 47\text{-}\mu F \ standard \ value)$$

$$C_2 = \frac{1}{\pi(15 \ k\Omega)(5 \ Hz)} \quad \text{(Eq. 4-9)}$$
$$= 4.3 \ \mu F \quad (use \ 4.7\text{-}\mu F \ standard \ value)$$

and

$$C_3 = \frac{1}{2\pi(500 \ \Omega)(5 \ Hz)} \quad \text{(Eq. 4-3)}$$
$$= 64 \ \mu F \quad (use \ 100\text{-}\mu F \ standard \ value)$$

The completed circuit is shown in Figure 4-6.

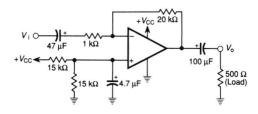

Fig. 4-6. Completed example circuit for a single supply biased inverting amplifier.

INVERTING SUMMING AMPLIFIER

The inverting summing amplifier biased for single supply voltage operation is a variation of the basic inverting amplifier. Figure 4-7 shows a 2-input summing amplifier circuit, whereby each input is AC-coupled, the output is AC-coupled to the load, and the amplifier's noninverting input is held at one-half the supply voltage. As a general rule, the cutoff frequencies of the input, bypass, and output RC networks are made the same. These are determined in the same manner as for the inverting amplifier.

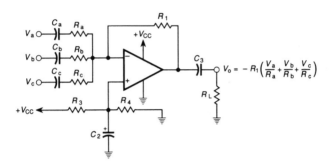

Fig. 4-7. Single supply biased summing amplifier.

DIFFERENCE AMPLIFIER

The difference amplifier powered by a single positive supply voltage is shown in Figure 4-8. When R_3 equals R_4, the DC voltage at the noninverting input is one-half the supply voltage. Like the conventional split supply difference amplifier discussed in Chapter 1, the design is greatly simplified by making $R_1 = R_2$, and these resistors are made equal to the desired input impedance of the amplifier.

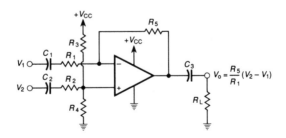

Fig. 4-8. Single supply biased difference amplifier.

The closed-loop differential gain is

$$A_D = \frac{R_5}{R_1} \qquad \text{(Eq. 4-12)}$$

while

$$A_D = \frac{R_3}{2R_2} \qquad \text{(Eq. 4-13)}$$

must hold. This in turn requires that

$$R_3 = 2R_5 \qquad \text{(Eq. 4-14)}$$

The AC output voltage of the difference amplifier is

$$V_o = \frac{R_5}{R_1}(V_2 - V_1) \qquad \text{(Eq. 4-15)}$$

The cutoff frequencies of the two input RC networks are given as

$$f_1 = \frac{1}{2\pi R_1 C_1} \qquad \text{(Eq. 4-16)}$$

and

$$f_2 = \frac{1}{2\pi(R_2 + R_3 \| R_4)C_2} \qquad \text{(Eq. 4-17)}$$

where

$$R_3 \| R_4 = \frac{R_3 R_4}{R_3 + R_4}$$

The cutoff frequency for the R_L-C_3 output network is selected using Equation 4-3. In practice, all three cutoff frequencies are made equal to provide the best common-mode rejection.

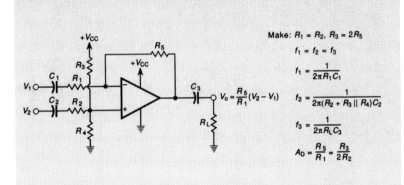

EXAMPLE 4-3

Determine the component values for the difference amplifier of Figure 4-8 driving a 1-kΩ load. The circuit is to have an input impedance of 10 kΩ for both inputs and a differential gain of 15 for frequencies above 100 Hz.

For an input impedance of 10 kΩ, then $R_1 = R_2 = 10$ kΩ. Resistors R_5 and R_3 are then determined from Equations 4-12 and 4-14

$$R_5 = (15)(10 \ k\Omega)$$
$$= 150 \ k\Omega \quad \text{(Eq. 4-12)}$$

and

$$R_3 = (2)(15)(150 \ k\Omega)$$
$$= 300 \ k\Omega \quad \text{(Eq. 4-14)}$$

which also requires that $R_3 = R_4$

$$R_4 = 300 \ k\Omega$$

For the closed-loop differential gain to be constant above 100 Hz, the cutoff frequencies of all three RC networks are made equal for the best common-mode rejection and set to approximately one-tenth 100 Hz, or 10 Hz. The capacitor value for each RC network is calculated

$$C_1 = \frac{1}{2\pi(10 \ k\Omega)(10 \ Hz)}$$ (Eq. 4-16)
$$= 1.6 \ \mu F \quad (use \ 2.2\text{-}\mu F \ standard \ value)$$

$$C_2 = \frac{1}{2\pi(10 \ k\Omega + 300 \ k\Omega \| 300 \ k\Omega)(10 \ Hz)}$$ (Eq. 4-17)
$$= 0.0995 \ \mu F \quad (use \ 0.1\text{-}\mu F \ standard \ value)$$

and

$$C_3 = \frac{1}{2\pi(1 \ k\Omega)(10 \ Hz)}$$ (Eq. 4-3)
$$= 15.9 \ \mu F \quad (use \ 15\text{-}\mu F \ standard \ value)$$

The completed circuit is shown in Figure 4-9.

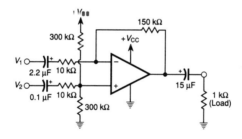

Fig. 4-9. Completed example circuit for a single supply biased difference amplifier.

Chapter 5

The Norton Amplifier

INTRODUCTION AND OBJECTIVES

Chapter 4 discusses the operation of conventional op-amps from a single supply voltage. In this chapter the Norton amplifier is presented, whose design specifically lends itself for single supply operation. Unlike a conventional op-amp that amplifies the difference in voltage between its inputs, the Norton amplifier produces an output voltage that is proportional to the difference in the currents flowing into its two input terminals.

At the completion of this chapter, you will be able to design and predict the operation of a Norton amplifier, such as the LM3900, using a single supply voltage in the following AC-coupled circuits:

- voltage follower
- inverting and noninverting amplifiers
- inverting and noninverting summing amplifiers
- difference amplifier

THE NORTON AMPLIFIER

A type of op-amp that is designed specifically for single supply operation is the *Norton amplifier* or *current differencing amplifier* (CDA). This type of device is very different from the conventional op-amp, as it was developed to provide a low-cost amplifier of modest performance that would work well from a single-ended power supply over a wide range of voltages with little change in its open-loop characteristics. As a tradeoff, some of its parameters are somewhat inferior to many general-purpose op-amps, such as the 741. Table 5-1 compares the LM3900 Norton amplifier to the 741 op-amp.

Table 5-1. Comparison of Typical Parameters of the LM3900 Norton Amplifier to the 741 Op-Amp

PARAMETER	741	LM3900
Supply Voltage Range	+5 − +15 V	+4 − +36 V
Open-Loop Gain	200,000	2,800
Unity Gain Frequency	1.0 MHz	2.0 MHz
Input Bias Current	80 nA	30 nA
Input Impedance	2 MΩ	1 MΩ
Output Impedance	75 Ω	8 kΩ
Slew Rate	0.5 V/μs	0.5 V/μs
PSRR	90 dB	70 dB

Unlike a conventional op-amp that amplifies the difference in voltage between its inputs, the Norton amplifier produces an output voltage that is proportional to the difference of the currents flowing into its input terminals. To distinguish itself from the conventional op-amp, the schematic symbol for the Norton amplifier includes a current source symbol (Fig. 5-1).

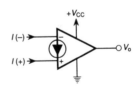

Fig. 5-1. Norton amplifier schematic symbol.

BIASING THE NORTON AMPLIFIER

Since the Norton amplifier is normally used to amplify AC signals, its output is usually biased to a DC level equal to one-half the power supply voltage. As with the bipolar op-amp, this output level permits an AC output signal to swing equally in either direction before clipping.

Figure 5-2 shows the basic biasing arrangement for the Norton amplifier, known as *current mirror biasing*. The DC output voltage is

$$V_o = \left(\frac{R_2}{R_1}\right) V_{CC} \quad \text{(Eq. 5-1)}$$

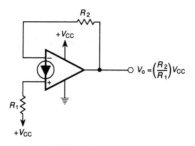

Fig. 5-2. Current mirror biasing.

For the DC output to be one-half the supply voltage, then

$$R_2 = \frac{R_1}{2} \quad \text{(Eq. 5-2)}$$

EXAMPLE 5-1

The basic current mirror bias circuit of Figure 5-2 using a Norton amplifier is biased with $R_1 = 390$ kΩ and $R_2 = 220$ kΩ. The DC quiescent output voltage for a supply voltage of 15 V is

$$V_o = \left(\frac{220 \text{ k}\Omega}{390 \text{ k}\Omega}\right) \times (15 \text{ V}) \quad \text{(Eq. 5-1)}$$

$$= 8.46 \text{ V}$$

BIASING WITH LOW SUPPLY VOLTAGES

If the supply voltage is less than 7 V, then Equation 5-1 must be changed to account for several internal diode drops (0.7 V), such that

$$R_1 = \left(\frac{V_{CC} - 0.7}{V_o - 0.7}\right) R_2 \quad \text{(Eq. 5-3)}$$

Since Norton amplifiers are designed to operate from a single-ended power supply and require coupling capacitors, such circuits will be capable then of amplifying AC signals only.

NORTON INVERTING AMPLIFIER

Figure 5-3 shows a Norton inverting amplifier. The closed-loop voltage gain is

$$A_{CL} = -\frac{R_2}{R_3} \qquad \text{(Eq. 5-4)}$$

The minus sign in Equation 5-4 indicates that the output voltage's polarity is opposite that of the input voltage and represents a 180° phase shift. In addition, since the closed-loop voltage gain is simply the ratio of two resistance values, the closed-loop gain can be made less than 1 (R_3 larger than R_2), equal to 1 (R_2 equal to R_3), or greater than 1 (R_2 greater than R_3).

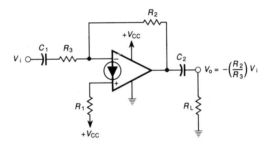

Fig. 5-3. Norton inverting amplifier.

The cutoff frequency of the input RC network is

$$f_1 = \frac{1}{2\pi R_3 C_1} \qquad \text{(Eq. 5-5)}$$

while the cutoff frequency of the output RC network is

$$f_2 = \frac{1}{2\pi R_L C_2} \qquad \text{(Eq. 5-6)}$$

These two frequencies are made equal and set approximately one-tenth the lowest frequency of the signal to be amplified. The value of R_3 equals the approximate input impedance of this stage, while the output impedance is very low, like the conventional inverting op-amp stage.

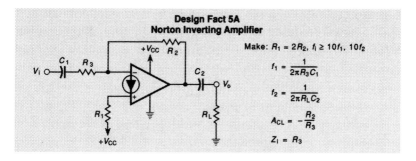

Design Fact 5A
Norton Inverting Amplifier

Make: $R_1 = 2R_2$, $f_1 \geq 10f_1$, $10f_2$

$f_1 = \dfrac{1}{2\pi R_3 C_1}$

$f_2 = \dfrac{1}{2\pi R_L C_2}$

$A_{CL} = -\dfrac{R_2}{R_3}$

$Z_i = R_3$

EXAMPLE 5-2

For the Norton inverting amplifier circuit of Figure 5-4, determine:

(a) The quiescent DC output voltage.
(b) The closed-loop AC voltage gain.
(c) The low-frequency response.
(d) The input impedance of the stage.

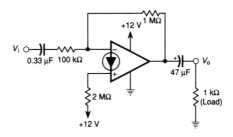

Fig. 5-4. Norton amplifier circuit example.

The quiescent DC output voltage is

$$V_o = \left(\dfrac{1\ M\Omega}{2\ M\Omega}\right) \times (12\ V)$$ (Eq. 5-1)

$$= 6\ V$$

and the closed-loop voltage gain is

$$A_{CL} = -\frac{1\ M\Omega}{100\ k\Omega}$$
$$= -10 \qquad \text{(Eq. 5-4)}$$

The cutoff frequencies for the input and output RC networks respectively are

$$f_1 = \frac{1}{2\pi(100\ k\Omega)(0.33\ \mu F)} \qquad \text{(Eq. 5-5)}$$
$$= 4.8\ Hz$$

$$f_2 = \frac{1}{2\pi(1\ k\Omega)(47\ \mu F)} \qquad \text{(Eq. 5-6)}$$
$$= 3.4\ Hz$$

The circuit will have a closed-loop voltage gain of 10 for input frequencies above approximately 10 times the dominant frequency (4.8 Hz) or 48 Hz.

Finally, the input impedance is approximately equal to the input resistance R_3 or 100 kΩ.

NORTON NONINVERTING AMPLIFIER

Figure 5-5 shows the circuit for a noninverting amplifier. This circuit is similar to the Norton inverting amplifier of Figure 5-3 except that R_3 and C_1 have been moved from the inverting input to the noninverting input. The closed-loop voltage gain is

$$A_{CL} = \frac{R_2}{R_3} \qquad \text{(Eq. 5-7)}$$

while the input impedance is approximately the input resistor R_3. These are the same for the inverting amplifier circuit.

Unlike the conventional bipolar noninverting amplifier which always has a closed-loop voltage gain greater than 1, the closed-loop voltage gain for the noninverting Norton circuit is simply the ratio of two resistance values. The closed-loop gain can then be made less than 1 (R_3 larger than R_2), equal to 1 (R_2 equal to R_3), or greater than 1 (R_2 larger than R_3). The cutoff frequencies for the three RC

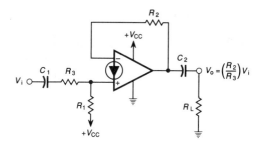

Fig. 5-5. Norton noninverting amplifier.

networks are determined in the same manner as for the inverting amplifier.

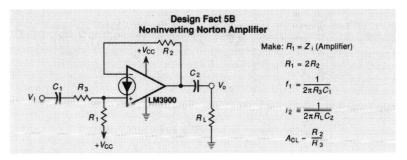

NORTON VOLTAGE FOLLOWER

By making $R_2 = R_3$ in the noninverting amplifier circuit of Figure 5-5, we have a unity-gain noninverting amplifier or voltage follower, shown in Figure 5-6. In order to achieve a high input impedance, R_3 should then be chosen as high as possible.

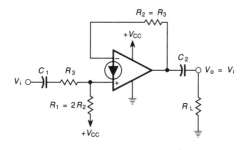

Fig. 5-6. Norton voltage follower.

NORTON SUMMING AMPLIFIERS

Summing amplifiers using Norton amplifiers are available in both inverting and noninverting configurations. Figure 5-7 shows a noninverting summing amplifier with three inputs. The AC output is easily obtained by considering each input as the input to a separate Norton noninverting amplifier and then combining the results. For the three-input circuit shown, the AC output voltage is

$$V_o = R_F \left(\frac{V_1}{R_1} + \frac{V_2}{R_2} + \frac{V_3}{R_3} \right) \quad \text{(Eq. 5-8a)}$$

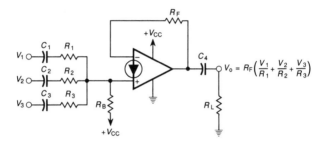

Fig. 5-7. Norton noninverting summing amplifier.

If the circuit is allowed to be simplified so that all three input resistors are made equal to R_1, the output voltage is then

$$V_o = \frac{R_F}{R_1} (V_1 + V_2 + V_3) \quad \text{(Eq. 5-8b)}$$

In both cases, the DC bias level at the Norton amplifier's output equals

$$V_o = \left(\frac{R_F}{R_B} \right) V_{CC} \quad \text{(Eq. 5-9)}$$

while the input impedance seen by each input signal source is approximately equal to the value of its corresponding series resistor

$$Z_1(in) = R_1 \quad \text{(Eq. 5-10a)}$$

$$Z_2(in) = R_2 \quad \text{(Eq. 5-10b)}$$

$$Z_3(in) = R_3 \qquad \text{(Eq. 5-10c)}$$

If we move the input circuit now to the Norton amplifier's inverting input as shown in Figure 5-8, we have the corresponding three-input Norton *inverting* summing amplifier. The analysis and output voltage equation are identical to that for the noninverting configuration, except that there is now a minus sign to reflect the change in output polarity

$$V_o = -R_F \left(\frac{V_1}{R_1} + \frac{V_2}{R_2} + \frac{V_3}{R_3} \right) \qquad \text{(Eq. 5-11a)}$$

or

$$V_o = -R_F \left(\frac{V_1}{R_1} + \frac{V_2}{R_2} + \frac{V_3}{R_3} \right) \qquad \text{(Eq. 5-11b)}$$

if all three input resistors are made equal to R_1.

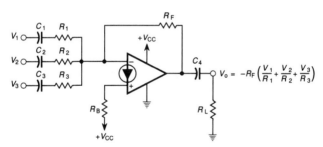

Fig. 5-8. Norton inverting summing amplifier.

Like the noninverting summing amplifier, the DC bias level at the Norton amplifier's output is given by Equation 5-9 while the input impedance seen by each input signal source is approximately equal to the value of its corresponding series resistor.

NORTON DIFFERENCE AMPLIFIER

The Norton difference amplifier, unlike its bipolar op-amp counterpart, does not require two pairs of matched resistors to set the closed-loop differential gain. Figure 5-9 shows the Norton difference amplifier using current mirror biasing. When both input voltages are equal, the DC quiescent output voltage is given by Equation 5-9.

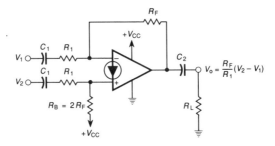

Fig. 5-9. Norton difference amplifier.

The differential gain is simply the ratio with an AC output voltage equal to

$$A_D = \frac{R_F}{R_1} \qquad \text{(Eq. 5-12)}$$

$$V_o = \frac{R_F}{R_1}(V_2 - V_1) \qquad \text{(Eq. 5-13)}$$

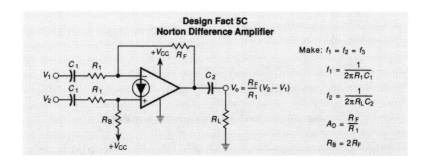

EXAMPLE 5-3

Determine the necessary component values for the Norton difference amplifier circuit of Figure 5-9 driving a 1.5-kΩ load. The circuit is to have a differential gain of 20 for frequencies above 100 Hz and an input impedance of 10 kΩ for both inputs.

For an input impedance of 10 kΩ, then $R_1 = R_2 = 10$ kΩ. For a differential gain of 20

$$R_F = (20)(10 \text{ k}\Omega)$$
$$= 200 \text{ k}\Omega \quad \text{(Eq. 5-12)}$$

To bias the quiescent DC output at one-half the supply voltage,

$$R_B = (2)(200 \text{ k}\Omega)$$
$$= 400 \text{ k}\Omega \quad (use\ 390\text{-k}\Omega\ standard\ value) \quad \text{(Eq. 5-9)}$$

The cutoff frequencies for all three RC networks should be set at approximately one-tenth 100 Hz or 10 Hz

$$C_1 = \frac{1}{2\pi(10 \text{ k}\Omega)(10 \text{ Hz})} \quad \text{(Eq. 5-5)}$$
$$= 1.6 \text{ }\mu F \quad (use\ 2.2\text{-}\mu F\ standard\ value)$$

$$C_2 = \frac{1}{2\pi(1.5 \text{ k}\Omega)(10 \text{ Hz})} \quad \text{(Eq. 5-6)}$$
$$= 10.6 \text{ }\mu F \quad (use\ 10\text{-}\mu F\ standard\ value)$$

The completed circuit is shown in Figure 5-10.

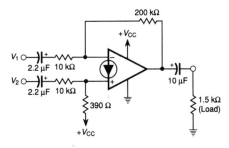

Fig. 5-10. Completed example circuit for a Norton difference amplifier.

Besides the AC amplifiers just discussed, other circuits (such as active filters and waveform generators) are possible with op-amps and Norton amplifiers powered from a single-ended supply.

Chapter 6

Nonlinear Signal Processing Circuits

INTRODUCTION AND OBJECTIVES

Very often, signals must be processed in other ways besides linear amplification. The resultant signal is often then used to either control another process, or transformed to allow easy measurement. This chapter discusses those signal processing circuits that are nonlinear. Here, the output signal is not a linear function of the input signal, as is the case with amplifiers. Nonlinear circuits include: precision rectifiers, comparators, peak detectors, sample-and-hold amplifiers, and logarithmic amplifiers.

At the completion of this chapter, you will be able to:

- Describe the operation and advantages of precision half- and full-wave rectifiers.
- Describe the operation of inverting, noninverting, and window comparator circuits, and explain what op-amp parameters should be considered.
- Explain the effect of noise on simple comparator circuits and how this may be eliminated.
- Describe the operation of peak detectors and sample-and-hold circuits.
- Describe the operation and limitations of logarithmic-type amplifiers and how they are used to multiply and divide two signals.

PRECISION RECTIFIERS

When a diode is used in half- and full-wave rectifiers, its nonlinear characteristics tend to distort the output waveform at low signal levels. Since silicon diodes must be forward biased to about 0.7 V and germanium diodes to about 0.3 V before conduction begins, neither of these devices alone are suitable for the rectification of small signal levels below several volts. The following subsections discuss how precise rectification can be accomplished with op-amps.

Half-Wave Rectifier

By incorporating such conventional diodes as the feedback element in an op-amp circuit, as shown in Figure 6-1a, it is possible to eliminate the effects of forward voltage drop of the diode on the output of the rectifier. Unlike most op-amp circuits, the output signal here is not taken from the output of the op-amp itself, but instead from the junction of R_1 and D_1. When the input signal is positive, D_2 is forward biased and all the feedback current flows through D_2. The output of the circuit at this point is zero because D_1 is reverse biased (Fig. 6-1b). When the input signal goes negative, D_2 is reverse

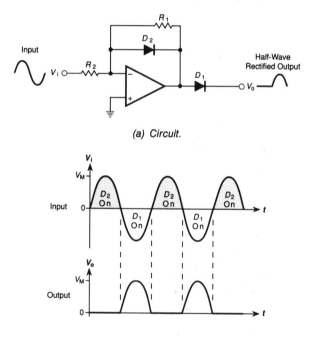

Fig. 6-1. Precision half-wave rectifier.

biased, and because D_1 is now forward biased, the feedback current flows through both D_1 and R_1 in series.

A signal then appears at the output across R_1. When the input level goes negative, the output of the op-amp jumps to the positive level required to cause D_1 to conduct. This effectively eliminates any diode drop from the output signal. Once D_1 begins to conduct, the circuit behaves as a conventional inverting amplifier. If $R_1 = R_2$, the closed-loop voltage gain is unity. The output signal then is nearly an exact half-wave rectified version of the input signal, but without the undesirable effects due to the diode drop. For this reason, this circuit is referred to as a *precision half-wave rectifier*. If half-wave rectification of the negative portion of the input signal is required, the polarity of the diodes are reversed.

Full-Wave Rectifier

The precision half-wave rectifier circuit of Figure 6-1a forms the basis for the *precision full-wave rectifier* of Figure 6-2a. It is composed of a precision half-wave rectifier followed by a 2-input summing amplifier.

One input to the summing amplifier is the signal to be rectified. The other input to the summing amplifier is the output from the precision half-wave rectifier. The summing amplifier then produces an output signal with two components. One component is the same level as the input signal and is inverted. The other component is due to the half-wave rectified output which is then inverted and amplified by a factor of two. The instantaneous sum of these two signal components is the full-wave rectified version of the input signal. Figure 6-2b shows how these signals combine to form the full-wave rectified sine-wave output.

Another precision full-wave rectifier circuit is shown in Figure 6-3. The first half of the circuit produces two complimentary half-wave outputs. The second half (A_2) is a difference amplifier.

THE PEAK DETECTOR

A *peak detector* is a circuit that will note and remember the peak positive or negative voltage of an input signal for an infinite period of time until it is reset. Such circuits are frequently used to measure the peak values of nonsinusoidal waveforms.

Figure 6-4 shows the circuit for a simple positive peak detector. The source signal charges the capacitor through the forward biased diode at a rate limited by the output impedance of the source. The capacitor will charge to the highest source voltage (less one diode

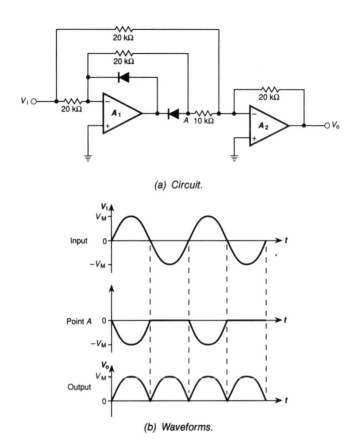

Fig. 6-2. Precision full-wave rectifier made from a precision half-wave rectifier and a 2-input summing amplifier.

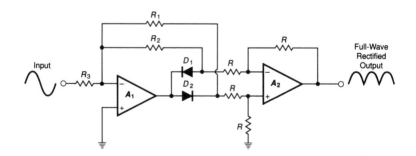

Fig. 6-3. Precision full-wave rectifier made from a difference amplifier.

drop) but cannot discharge itself except through capacitor leakage or very high input impedance of the op-amp. Since the op-amp is connected as a voltage follower, its output voltage equals the positive peak value of the applied signal and will remain that way for relatively long periods.

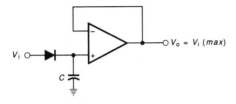

Fig. 6-4. Basic peak detector.

Figure 6-5 shows an improved peak detector that buffers the signal source from the capacitor. Op-amp A_1 presents a high-impedance load to the signal source while A_2 serves as a buffer between the capacitor and the load. With this circuit, the DC output voltage (V_o) at any given time equals the peak capacitor voltage. This action assumes that the capacitor has had sufficient time to charge. The relatively low output impedance of A_1 serves to improve the circuit's response time for high-impedance sources.

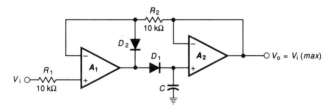

Fig. 6-5. Improved peak detector circuit using a buffer amplifier.

If the level of the input signal becomes greater than the stored capacitor voltage, the capacitor will then charge towards the higher input level. On the other hand, if the input level now drops below the stored capacitor voltage, the output remains at the former value.

Resistor R_1 is made equal to R_2 in order to minimize the effect of the op-amp's offset voltage. Since A_1 is used to drive a capacitive load and is operated with unity gain, it is important to be sure that any required frequency compensation is provided in order to avoid instability (i.e., oscillations). In addition, A_2 is usually a JFET-input op-amp for its high input impedance.

COMPARATORS

When processing analog signals, it is often necessary to know if an instantaneous signal level is more positive or negative than another signal, or if a given signal is inside or outside the range of two fixed voltage levels. On the other hand, it is sometimes required to determine when the signal's polarity changes. A class of circuits that is capable of making these determinations is the *comparator*.

Sometimes called a *level detector*, the comparator uses the op-amp in its open-loop mode to compare the instantaneous input signal level against a fixed reference voltage.

Basic Comparators

Figure 6-6a shows a *noninverting comparator*. The input signal (V_i) is connected to the noninverting input and a fixed DC reference voltage (V_{REF}) is connected to the inverting input. The reference can be either a positive or negative voltage. Whenever V_i is greater than V_{REF}, even by as little as 1 mV, the high open-loop gain of the op-amp forces the output voltage to its positive saturation voltage, V_{+SAT}, which is typically 1.4 V to 2 V less than the corresponding supply voltage. On the other hand, whenever V_i is less than V_{REF}, the output voltage then swings to its negative saturation voltage of V_{-SAT}.

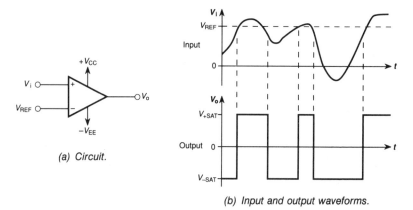

Fig. 6-6. Noninverting comparator.

The ideal switching level is when $V_i - V_{REF} = 0$. However, this action is affected somewhat by the op-amp's input offset voltage, which should be compensated for when exact switching is required. The circuit of Figure 6-6a is often called a *noninverting comparator*, as the output goes positive when the input signal is greater than the reference-voltage reference (Fig. 6-6b). In summary,

$$V_o = V_{+SAT} \text{ when } V_i > V_{REF}$$
$$= V_{-SAT} \text{ when } V_i < V_{REF}$$
$$= 0 \text{ when } V_i = V_{REF}$$

(Eq. 6-1)

The reference voltage can be generated in a variety of ways. Figure 6-7a shows a fixed resistive voltage divider whose source voltage is the same regulated voltage used to power the op-amp.

On the other hand, Figure 6-7b shows the use of a zener diode to establish the reference voltage. Here the reference voltage equals the zener voltage of the diode used.

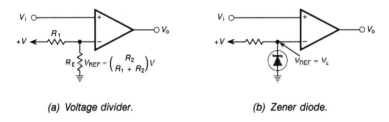

(a) Voltage divider. (b) Zener diode.

Fig. 6-7. Common methods for establishing a reference voltage.

If the reference voltage is made zero by grounding the op-amp's inverting input, the resultant circuit of Figure 6-8a, is called a *zero crossing detector*. The output switches from one extreme to the other when the input signal level crosses zero volts. This same circuit can also be used as a *polarity detector* since the output is positive when the input is positive, and negative when the input is negative with respect to ground.

As shown by the waveforms in Figure 6-8b, one popular use for the zero reference circuit is converting a sine wave into a square wave.

Figure 6-9a shows an *inverting comparator* obtained by reversing the inputs of the noninverting comparator, which has the opposite action as the noninverting circuit. Whenever V_i at the inverting input is greater than V_{REF}, the output swings towards V_{-SAT}. Whenever V_i is less than V_{REF}, the output voltage then swings towards V_{+SAT} (Fig. 6-9b). In summary,

$$V_o = V_{-SAT} \quad \text{when } V_i > V_{REF}$$
$$= V_{+SAT} \quad \text{when } V_i < V_{REF} \quad \text{(Eq. 6-2)}$$
$$= 0 \quad \text{when } V_i = V_{REF}$$

To be useful, comparator circuits must be interfaced with other circuits. This interface may be as simple as the LED driver of Figure 6-10a which indicates the state of the output level. When interfaced with a noninverting comparator as shown, the LED is lit when V_i is greater than V_{REF}.

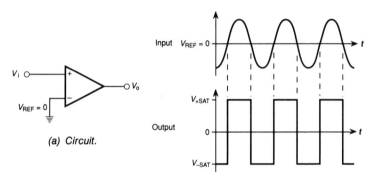

Fig. 6-8. Zero crossing detector.

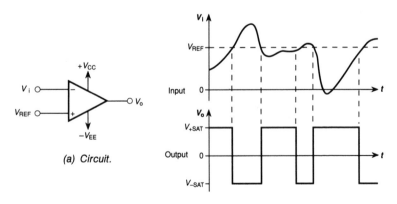

Fig. 6-9. Inverting comparator.

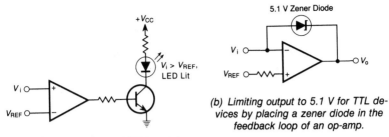

Fig. 6-10. Comparator interfacing circuits.

On the other hand, comparators frequently drive TTL and CMOS digital circuits where the logic 0 and 1 levels are well defined, and the digital device may be destroyed, if exceeded. Consequently, it is undesirable for comparators to produce large positive and negative output voltage levels. For example, TTL devices require that the maximum positive voltage representing a logic 1 is typically +5 V, while logic 0 is at ground. If a comparator operating from a 12-V split supply is used, the output will swing from −12 V to +12 V, depending on the relationship of V_i to V_{REF}.

Figure 6-10b shows a zener diode as a feedback element to limit the positive output voltage to the zener voltage. A 5.1-V zener diode would be used for TTL voltage levels.

Noise Effects and Comparators with Hysteresis

Suppose that the levels of the two comparator inputs are very close to each other and there is a small noise level on one of these inputs. If the noise level is large enough, as shown in Figure 6-11, the output of a conventional comparator will then switch back and forth between V_{+SAT} and V_{-SAT} each and every time the noisy input signal exceeds or drops below the reference voltage. These rapid changes due to noise are unwanted transitions and are sometimes termed *chatter*.

To prevent noise from causing false comparator chattering, positive feedback is added to provide a *dead zone* or noise immunity over which the comparator will not respond. Such a circuit, called a *Schmitt trigger*, is shown in Figure 6-12a. The voltage at the noninverting input determines the positive-going level at which V_i will cause the comparator's output to switch to V_{-SAT}. This positive-going level is the *upper threshold voltage* (V_{UT}) and is found

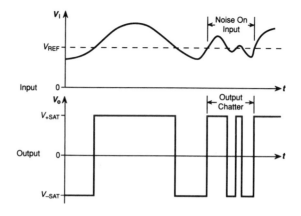

Fig. 6-11. Effect of input noise on the output of a comparator.

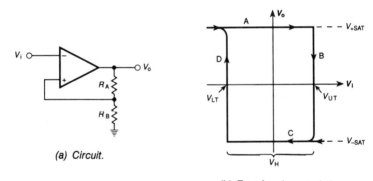

(a) Circuit.

(b) Transfer characteristic.

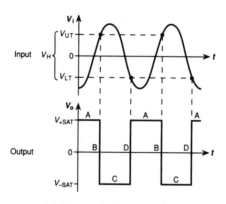

(c) Input and output waveforms.

Fig. 6-12. Inverting comparator with hysteresis.

$$V_{UT} = \left(\frac{R_B}{R_A + R_B}\right) V_{+SAT} \qquad \text{(Eq. 6-3)}$$

When V_i is less than V_{UT}, the voltage on the noninverting input is more positive than V_i and the output equals V_{+SAT}. When V_i exceeds V_{UT}, the output switches to V_{-SAT}. When the output voltage equals V_{-SAT}, the voltage across R_B is called the *lower threshold voltage* (V_{LT}). The input voltage then must drop below V_{LT} before the output will switch to V_{+SAT} when

$$V_{LT} = \left(\frac{R_B}{R_A + R_B}\right) V_{-SAT} \qquad \text{(Eq. 6-4)}$$

The width of dead zone, which is centered about zero volts, is called the *hysteresis voltage* (V_H) and is the difference between the two threshold levels

$$V_H = V_{UT} - V_{LT} \qquad \text{(Eq. 6-5)}$$

Usually the upper and lower threshold voltages are equal except for polarity. For voltage levels within the dead zone from V_{LT} to V_{UT}, the comparator does not respond. Only when the input levels exceed these thresholds does the output change.

The effect of hysteresis is shown in Figure 6-12b, while Figure 6-12c illustrates the effect of hysteresis on a sine-wave signal.

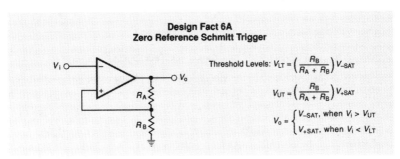

EXAMPLE 6-1

The op-amp used in the Schmitt trigger circuit of Figure 6-13 using a +12 V supply has a saturation voltage of +10.5 V.

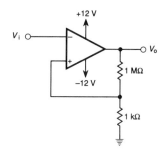

Fig. 6-13. Noninverting comparator example circuit.

The upper threshold voltage is

$$V_{UT} = \frac{1\ k\Omega}{(1\ M\Omega + 1k\Omega)}(10.5\ V) \quad \text{(Eq. 6-3)}$$
$$= 10.5\ mV$$

while the the lower threshold voltage is

$$V_{LT} = \frac{1\ k\Omega}{(1\ M\Omega + 1k\Omega)}(10.5\ V) \quad \text{(Eq. 6-4)}$$
$$= -10.5\ mV$$

The threshold voltage is then

$$V_H = 10.5\ mV - (-10.5\ mV) \quad \text{(Eq. 6-5)}$$
$$= 21\ mV$$

Upper and Lower Switching Levels for Nonzero References

In many situations however, it is desired that the reference be some value other than zero while still having some amount of noise immunity. If the lower end of R_B of the basic Schmitt trigger is tied to a DC reference voltage, as shown by the inverting comparator circuit of Figure 6-14a, the upper and lower threshold voltages respectively are now

$$V_{UT} = \frac{R_B(V_{+SAT} - V_{REF})}{R_A + R_B} + V_{REF} \quad \text{(Eq. 6-6a)}$$

and

$$V_{LT} = \frac{R_B(V_{-SAT} - V_{REF})}{R_A + R_B} + V_{REF} \qquad \text{(Eq. 6-6b)}$$

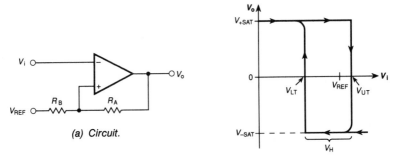

(a) Circuit.

(b) Transfer characteristic.

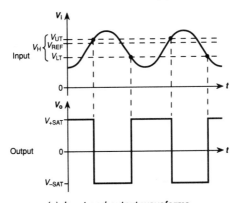

(c) Input and output waveforms.

Fig. 6-14. Inverting comparator with hysteresis and nonzero voltage reference.

Figure 6-15a shows the corresponding circuit for a noninverting comparator with hysteresis. Here, the inputs to the op-amp are reversed from those of Figure 6-14a. For this circuit, the switching levels are given by

$$V_{UT} = \left(\frac{R_A + R_B}{R_A}\right)V_{REF} - \left(\frac{R_B}{R_A}\right)V_{+SAT} \qquad \text{(Eq. 6-7a)}$$

$$V_{LT} = \left(\frac{R_A + R_B}{R_A}\right)V_{REF} - \left(\frac{R_B}{R_A}\right)V_{-SAT} \qquad \text{(Eq. 6-7b)}$$

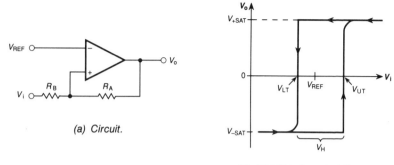

(a) Circuit.

(b) Transfer characteristic.

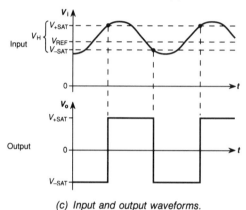

(c) Input and output waveforms.

Fig. 6-15. Noninverting comparator with hysteresis and nonzero voltage reference.

Window Comparators

A *window comparator*, also called a *double-ended comparator*, is a circuit that detects whether or not an input voltage is between two specified voltage limits, called a *window*. As shown in Figure 6-16a, this is normally accomplished by logically combining the outputs from both an inverting comparator and a noninverting comparator.

When the input level is greater than the upper reference voltage (V_{UL}) or less than the lower reference voltage (V_{LL}) of the window, the output of the circuit is at V_{+SAT}. If the level of the input voltage is in the window between V_{LL} and V_{UL}, the output voltage is zero. In summary

$$\begin{aligned} V_o &= 0 \quad \text{when } V_{LL} < V_i < V_{UL} \\ &= V_{+SAT} \quad \text{when } V_i < V_{LL} \text{ or } V_i > V_{UL} \end{aligned} \quad \text{(Eq. 6-8)}$$

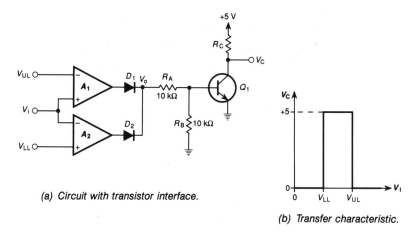

Fig. 6-16. Window comparator.

(a) Circuit with transistor interface.

(b) Transfer characteristic.

The transistor output driver allows for output level conversion. As shown, the transistor supply is +5 V. If the input signal is within the window, the high voltage at the transistor's base will turn Q_1 on, and the voltage at the collector (V_C) will be essentially zero. On the other hand, if the input is outside the window, the transistor is cut off so that the collector voltage is approximately +5 V. This action is graphed in Figure 6-16b. The collector resistor (R_C) is sized according to the transistor's maximum collector current rating, $I_C(max)$. Using a 5-V supply, this minimum value is

$$R_C(\min) = \frac{5\ V}{I_C(\max)} \qquad \text{(Eq. 6-9)}$$

This type of circuit is frequently used in alarm circuits having high and low trip points. For example, a thermistor could be used to monitor the temperature within an enclosure by producing an output voltage proportional to its resistance, which is dependent on the measured temperature. If the upper and lower reference voltage limits are chosen to correspond to voltage levels representing high and low temperature limits, the window comparator can be used to monitor when the measured temperature is outside a given range. When outside the prescribed range, the collector voltage is sufficient to drive a bell, siren, or some other type circuit.

EXAMPLE 6-2

For the window comparator circuit of Figure 6-17, determine the range of input voltages for which the LED will be lit.

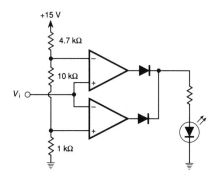

Fig. 6-17. Window comparator example circuit.

The upper and lower threshold limits are set by the resistor divider string. The upper threshold is

$$V_{UL} = \frac{(10\ k\Omega + 1\ k\Omega)}{(4.7\ k\Omega + 10\ k\Omega + 1\ k\Omega)}(15\ V)$$

$$= +10.51\ V$$

while the lower threshold is

$$V_{LL} = \frac{(1\ k\Omega)}{(4.7\ k\Omega + 10\ k\Omega + 1\ k\Omega)}(15\ V)$$

$$= +0.96\ V$$

The LED will then be lit whenever the input voltage is outside the window. For this circuit, the LED will be lit for input levels less than +0.96 V, as well as for those greater than +10.51 V.

Norton Amplifier Comparators

The Norton amplifier, or CDA, which is discussed in Chapter 5, is also capable of performing several comparator operations. The basic CDA comparator is shown in Figure 6-18. In terms of input voltage levels, the output voltage is

$$V_o = V_{+SAT} \quad \text{when } V_i > V_{REF}$$
$$ = 0 \quad \text{when } V_i = V_{REF}$$
(Eq. 6-10)

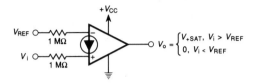

Fig. 6-18. Norton amplifier noninverting comparator.

Although the CDA is designed to be powered by a positive supply voltage, it can nevertheless be made to compare negative voltages. In the circuit of Figure 6-19, resistors R_1 and R_2 establish a *common-mode bias current*, allowing V_i and/or V_{REF} to be negative voltage levels.

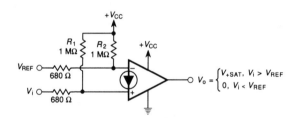

Fig. 6-19. Norton amplifier noninverting comparator biased to compare negative voltages when the circuit is powered by a single positive supply voltage.

A Norton window comparator circuit is shown in Figure 6-20. Resistors R_1 and R_4 are made equal, while the lower and upper threshold voltages are

$$V_{LT} = \left(\frac{R_3}{R_4}\right) V_{REF}$$
(Eq. 6-11a)

$$V_{UT} = \left(\frac{R_2}{R_1}\right) V_{REF}$$
(Eq. 6-11b)

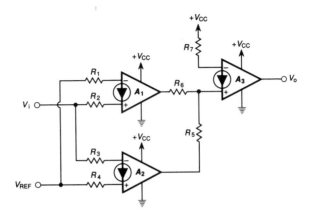

Fig. 6-20. Norton amplifier window comparator.

Both outputs are tied to a third CDA comparator through R_4 and R_5. For proper biasing, $R_4 = R_5 = R_6/2$.

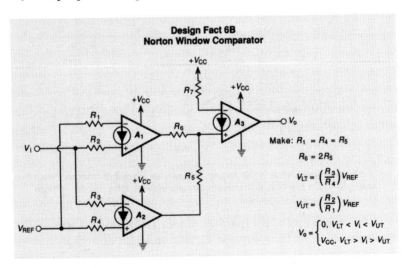

Limitations and Precautions

Several things must be kept in mind when selecting either an op-amp or CDA to be used as a comparator. First of all, the device must have a high slew rate so that the output will change as rapidly as possible. In addition, for high-speed operation, the unity-gain frequency should be as high as possible.

In order to obtain accurate switching levels with or without hysteresis, the device should also have a low input offset voltage, low input bias current, and high CMR ratings. Despite these desirable factors, a general-purpose device like the 741 is reliable for low-speed, less critical applications.

SAMPLE-AND-HOLD AMPLIFIER

Sample-and-hold amplifiers, sometimes abbreviated *S/H*, are used to (1) capture the value of a rapidly changing signal at a specific point in time and to (2) hold it steady. This is important during analog-to-digital conversion operations, as well as the storage of the outputs of a multiplexer between updates in data-distribution systems.

To understand sample-and-hold operation, first consider a circuit network having an input and output port, as well as a control line. When the control line is held at some high-level voltage, the output follows the input. When the control line is grounded, the output is held or *frozen* at its current value. The output of the S/H circuit will change only when the control line is taken to the high-level voltage (Fig. 6-21).

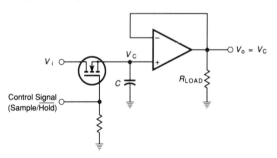

Fig. 6-21. Input and output waveforms illustrating the operation of a sample-and-hold amplifier.

Figure 6-22 shows one of many types of sample-and-hold circuits possible. The input signal is buffered from the capacitor by A_1, which provides a low-impedance source for charging and discharging the capacitor. Op-amp A_2 serves as a high-impedance load for the capacitor to buffer it from the actual load. The overall accuracy and response time of the circuit is improved by using V_o as the negative feedback signal to A_1 instead of merely connecting the output of A_1 back to its inverting input. This arrangement also compensates for common-mode and offset errors of the output voltage follower.

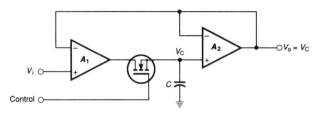

Fig. 6-22. Basic sample-and-hold amplifier circuit using a buffer amplifier.

LOGARITHMIC AND ANTILOGARITHMIC AMPLIFIERS

A *logarithmic amplifier*, or *log amp*, is a circuit that produces an output signal proportional to the logarithm of the input signal. Historically, logarithmic-type amplifiers were used in analog computers. Today, logarithmic amplifiers find many uses in circuits that include multipliers, dividers, RMS converters, and dynamic range compressors. Although this chapter discusses these circuits using op-amps, increased accuracy and flexibility are obtained with those integrated circuit devices that are specifically designed for these purposes.

The P-N Junction

The semiconductor p-n junction like those found in diodes and transistors is the key to all logarithmic amplifiers. Because of the increased dynamic range that can be obtained however, transistors are generally preferred for discrete logarithmic amplifier circuits. The transistor base-emitter voltage (V_{BE}) is a logarithmic function of its collector current (I_C)

$$V_{BE} = \frac{2.3\ kT}{q} \log_{10}(I_C/I_{RS}) \qquad \text{(Eq. 6-12)}$$

where:
    k = Boltzmann's constant, 1.38×10^{-23} J/K
    T = absolute temperature (K)
    q = charge on an electron, 1.6×10^{-19} C
    I_C = collector current (A)
    I_{RS} = theoretical reverse saturation current (A)

The theoretical reverse saturation current is typically 0.1 pA at 25°C. Furthermore at 25°C (298 K), Equation 6-12 reduces to

$$V_{BE} = 0.0585\ \log_{10}(I_C) + 0.761 \qquad \text{(Eq. 6-13)}$$

Logarithmic Amplifier

Figure 6-23 shows a logarithmic amplifier circuit, which is also called a *transdiode logarithmic converter*. Here, the feedback element is a grounded-base transistor, and is often a power transistor that reduces the series feedback resistance.

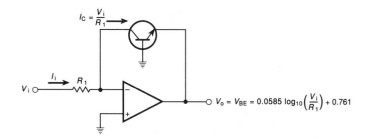

Fig. 6-23. Op-amp transistor logarithmic amplifier.

The transistor's collector current equals the input current of the op-amp. Since the base of the transistor is grounded, the output of the op-amp is equal to V_{BE}. Consequently, the output voltage is proportional to the logarithm of the input current, which in turn is proportional to the input voltage. Because of the virtual ground at the inverting input, the collector and base voltages of the transistor are equal.

For this circuit, Equation 6-13 is written as

$$V_{BE} = 0.0585 \log_{10} \left(V_i/R_1 \right) + 0.761 \quad \text{(Eq. 6-14)}$$

at 25°C. If negative-level inputs are required, then a *pnp* transistor is used so that the output will be positive.

The discrete transdiode amplifier has several negative characteristics. First, Equation 6-14 is temperature dependent. Secondly, the output changes only about 60 mV per decade, where 1 V/decade might be more desirable. These two primarily effect the scaling factor of 60 mV/decade. On the other hand, the current I_{RS} is also temperature sensitive and contributes a voltage offset in Equation 6-12 that is approximately 761 mV at 25°C.

All three of these problems can be compensated for with additional circuitry. However, commercially available integrated circuit logarithmic amplifier modules are designed to automatically compensate for these errors.

Antilogarithmic Amplifier

The antilogarithmic amplifier has an output voltage proportional to the inverse log, or antilogarithm of the input voltage. Like multiplication and division, as well as integration and differentiation being inverse mathematical operations, so are logarithm and antilogarithm inverse operations.

By interchanging the resistor and the transistor of the logarithmic amplifier circuit of Figure 6-23, the antilogarithmic amplifier of Figure 6-24 is formed. The output is proportional to the antilog or inverse logarithm of the input

$$V_o = -R_F I_{RS} e^{(V_i q/kT)} \qquad \text{(Eq. 6-15)}$$

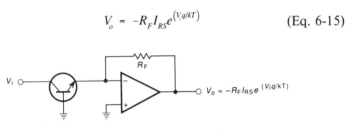

Fig. 6-24. Antilogarithmic amplifier.

Like the logarithmic amplifier, there are commercially available integrated-circuit antilogarithmic amplifier modules that are designed to offer improved characteristics.

Multiplication and Division

Discrete logarithmic and antilogarithmic amplifiers are used to build analog circuits capable of multiplying and dividing two analog voltages.

Figure 6-25 is a block diagram of an analog multiplier using two logarithmic amplifiers and a 2-input summing amplifier. The output of the summing amplifier equals the logarithm of the product of the two input signals (X and Y) and is equivalent to the sum of the individual logarithms of the two numbers

$$\log(XY) = \log(X) + \log(Y) \qquad \text{(Eq. 6-16)}$$

If the log (XY) signal is passed through an antilogarithmic amplifier, the output is equal to the product (XY) of the two signals

$$XY = \log^{-1}[\log(XY)] \qquad \text{(Eq. 6-17)}$$

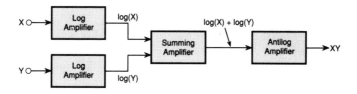

Fig. 6-25. Block diagram showing how to multiply two signals.

In a similar manner, as shown in block diagram form in Figure 6-26, we are able to divide one signal by another (i.e., X/Y). The output of the difference amplifier equals the logarithm of the quotient of the two input signals and is equivalent to the difference of the individual logarithms of the two numbers

$$\log(X/Y) = \log(X) - \log(Y) \qquad \text{(Eq. 6-18)}$$

If the log (X/Y) signal is passed through an antilogarithmic amplifier, the output is equal to the quotient (X/Y) of the two signals

$$X/Y = \log^{-1}[\log(X/Y)] \qquad \text{(Eq. 6-19)}$$

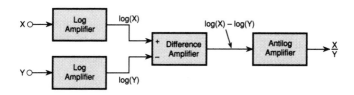

Fig. 6-26. Block diagram showing how to divide two signals.

117

Chapter 7

Voltage and Current Regulators

INTRODUCTION AND OBJECTIVES

A voltage regulator provides a constant DC voltage level to its load regardless of changes in source voltage or load current. A voltage regulator normally interfaces with an unregulated or poorly regulated DC power source and a load that requires a constant voltage source. The power supply and its need to produce a source of regulated voltage is perhaps the most important section of any active circuit design. Without a power supply, no active circuit could operate.

This chapter discusses those positive and negative linear regulators built around an op-amp. Depending on how the control element is connected, linear regulators are classified as either series or shunt. Besides regulating voltage, the op-amp can be also used to regulate current to a load.

At the completion of this chapter, you will be able to:

- Draw the diagram and explain the operation of a series regulator.
- Draw the diagram and explain the operation of a shunt regulator.
- Explain how to implement constant-current limiting and foldback current limiting as well as their advantages and disadvantages.
- Increase the output current of a regulator using a pass transistor.
- Discuss the operation of a current regulator or constant-current source.

SERIES REGULATORS

For low-current applications of 100 mA or less, op-amps are often used since these simple design and low-cost devices provide stable regulation of circuit parameters. For higher current applications requiring load currents greater than 100 mA, a series-pass transistor is added.

Basic Circuits

The circuit for a simple op-amp positive-voltage, low-current series regulator is shown in Figure 7-1. The op-amp functions as a voltage follower and supplies the load current directly. The voltage drop across a reverse-biased zener diode is the reference voltage input, $V_Z = V_{REF}$.

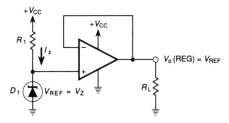

Fig. 7-1. Positive-voltage, low-current series regulator using a zener diode.

The unregulated DC source voltage (V_{CC}) serves to reverse bias the zener diode as well as to furnish the necessary positive supply voltage for the op-amp. Resistor R_1 is determined from

$$R_1 = \frac{V_{CC} - V_{REF}}{I_Z} \qquad \text{(Eq. 7-1)}$$

and sets the proper current level (I_Z) through the zener diode to give its rated zener voltage which is generally specified on the zener's data sheet. The power dissipation of R_1 is determined

$$P_{R1} = (V_{CC} - V_{REF})I_Z \qquad \text{(Eq. 7-2)}$$

In addition, the op-amp's negative supply pin is grounded.

The value of the output voltage must be kept several volts less than V_{CC}, which then prevents the op-amp output from saturating. Consequently, a sufficient amount of the supply voltage must be dropped across the regulator to ensure an adequate voltage regulation range.

This type of op-amp regulator is an improvement over that obtained from a simple zener-diode current limiting resistor circuit. With the addition of the high input resistance of the op-amp, the current through the zener diode is held essentially constant even though the load current may vary. Furthermore, the op-amp also provides a much lower source resistance for the load and a greater current capacity than that provided by the combination of a small zener diode and a resistor standing alone.

A regulated output higher than the zener voltage is achieved by connecting the op-amp as a noninverting amplifier, shown in Figure 7-2. Here, the unregulated input voltage supplies power to the op-amp as well as establishing the reference voltage across the zener diode (D_1). The operation of this circuit is very similar to the circuit in Figure 7-1, except that the reference voltage is compared with a fraction of the output voltage taken from a voltage divider. The regulated output voltage is

$$V_o(REG) = \left(1 + \frac{R_2}{R_3}\right) V_{REF} \qquad \text{(Eq. 7-3)}$$

Since most op-amps have rather limited current output capabilities, series regulator circuits are only suitable for applications which require small load currents. If the op-amp drives an NPN power transistor connected as an emitter follower and taking the feedback signal from the emitter, as shown in Figure 7-3a, the transistor behaves as though it is part of the output stage of the op-amp. The transistor gives this regulator a much greater output current capacity, now limited by the current and power ratings of the transistor and not those of the op-amp.

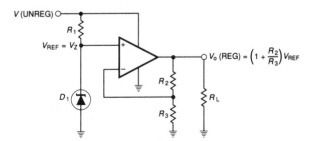

Fig. 7-2. Using a noninverting amplifier to generate a regulated output voltage greater than the zener reference.

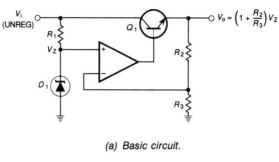

(a) Basic circuit.

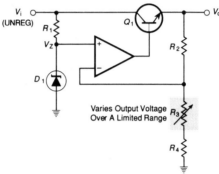

(b) Using a transistor to increase the output current of a regulator.

Fig. 7-3. Series-pass voltage regulator with adjustable output voltage.

The only calculations required are for the zener-diode reference (D_1 and R_1) and the specification of Q_1 [$I_C(max)$, $P_D(max)$, and $V_{CE}(max)$]. As a general rule, the zener diode chosen has a zener voltage that is approximately one-half the output voltage. The current flowing through R_2 and R_3 making up the voltage divider network must be greater than the input bias current flowing into the inverting input of the op-amp. In most cases, the current flowing through R_2 and R_3 can be set to about 1 mA since the input bias current is typically less than 100 nA. Resistors R_2 and R_3 are then determined from

$$R_2 = \frac{V_{CC} - V_{REF}}{1\ mA} \quad \text{(Eq. 7-4)}$$

and

$$R_3 = \frac{V_{REF}}{1\ mA} \quad \text{(Eq. 7-5)}$$

High-power regulators of this type often require that the transistor be mounted on a heat sink in order to increase its power handling capabilities. Figure 7-3b shows the same circuit, except that the regulated output voltage can be varied over a limited range with the potentiometer R_3.

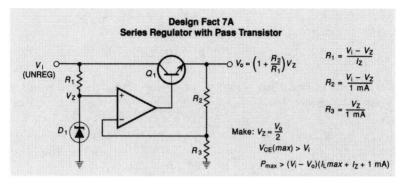

Design Fact 7A
Series Regulator with Pass Transistor

$V_o = \left(1 + \frac{R_2}{R_1}\right) V_z$

$R_1 = \frac{V_i - V_z}{I_z}$

$R_2 = \frac{V_i - V_z}{1 \text{ mA}}$

$R_3 = \frac{V_z}{1 \text{ mA}}$

Make: $V_z = \frac{V_o}{2}$

$V_{CE}(max) > V_i$

$P_{max} > (V_i - V_o)(I_{L\,max} + I_z + 1 \text{ mA})$

EXAMPLE 7-1

Using the variable positive-voltage regulator circuit of Figure 7-3b, determine the necessary component values so that, given an 18-V unregulated source, the output voltage is adjustable from 9 to 15 V with a maximum output current of 150 mA.

As a general rule, the zener-diode voltage is approximately one-half the output voltage. Since the output is to vary from 9 V to 15 V, any zener diode having a zener voltage from 4.5 V to 7.5 V will do. Here, assume that a 5.1-V zener diode is chosen. From the diode's data sheet, this voltage typically requires that 20 mA flow through the zener diode for 5.1 V. The value for R_1 is then determined on the basis of the source voltage (18 V) and the average output voltage (12 V)—that is, the average of 9 to 15 V

$$R_1 = \frac{18 \text{ V} - 12 \text{ V}}{20 \text{ mA}}$$
$$= 300 \text{ } \Omega \qquad \text{(Eq. 7-1)}$$

The average power dissipated by R_1 is then

$$P = (20 \text{ mA})(6 \text{ V})$$
$$= 120 \text{ mW} \qquad \text{(Eq. 7-2)}$$

so that a 300-Ω, 1/4-W standard resistor can be used.

The current flowing R_2, R_3, and R_4 is set at about 1 mA. When the output voltage is at 9 V, the wiper arm is at the top of the potentiometer R_3. The total voltage drop across R_3 and R_4 must be equal to the voltage reference of the zener diode (5.1 V), so that the total resistance is

$$R_3 + R_4 = \frac{5.1\ V}{1\ mA}$$
$$= 5{,}100\ \Omega$$

On the other hand, the average voltage drop across R_2 is 12 V − 5.1 V, or 6.9 V. Since 1 mA also flows through R_2

$$R_2 = \frac{6.9\ V}{1\ mA}$$
$$= 6{,}900\ \Omega \qquad (use\ 6.8\text{-}k\Omega\ standard\ value) \qquad \text{(Eq. 7-4)}$$

When the output is at its maximum of 15 V, the wiper arm is now at the bottom of R_3. The maximum current flowing through R_2, R_3, and R_4 is now

$$I = \frac{15\ V}{5.1\ k\Omega + 6.8\ k\Omega}$$
$$= 1.3\ mA$$

so that R_4 is found from

$$R_4 = \frac{5.1\ V}{1.3\ mA}$$
$$= 3{,}923\ \Omega \qquad (use\ 3.9\text{-}k\Omega\ standard\ value)$$

Therefore

$$R_3 = 5.1\ k\Omega - 3.9\ k\Omega$$
$$= 1.2\ k\Omega$$

for which a 1-kΩ potentiometer can be used without much loss of accuracy.

When Q_1 is at cutoff, $V_{CE}(max) = 18$ V. The maximum collector current is

$$I_C(max) = 150\ mA + 20\ mA + 1.3\ mA$$
$$= 171.3\ mA$$

The maximum power dissipation across the collector-emitter junction is

$$P_D(\text{max}) = (18\ V - 9\ V)(171.3\ mA)$$
$$= 1.54\ W$$

Any transistor having at least the minimum values determined for $V_{CE}(\text{max})$, $I_C(\text{max})$, and $P_D(\text{max})$ will work. The final circuit is shown in Figure 7-4.

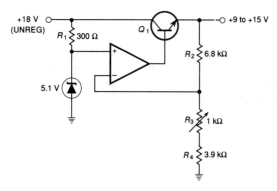

Fig. 7-4. Completed example circuit for an adjustable series-pass regulator.

Constant-Current Limiting

If there is a short circuit or overload on the output of a voltage regulator circuit, it is nice to have some form of current limiting to prevent damage to the regulator components. Figure 7-5 shows a simple means of limiting output current during a short circuit or overload.

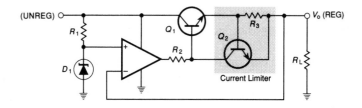

Fig. 7-5. Series-pass voltage regulator with current limiting.

When the load current reaches a predetermined value, the voltage across resistor R_3 reaches approximately 0.7 V. This then is sufficient to forward bias Q_2 and divert all the current away from the base of the series-pass transistor (Q_1) to the collector of Q_2. This action then puts an upper limit on the current which can be delivered by the regulator to slightly more than the value necessary to cause a 0.7-V drop across R_3.

$$I_L(\max) = \frac{0.7\ V}{R_3} \qquad \text{(Eq. 7-6)}$$

As long as Q_2 is activated by some minimum current level, the base-emitter voltage (V_{BE}) will remain constant across R_3. This in turn clamps the current through R_3 to some maximum value. If the load should suddenly demand additional current, it will then be limited by R_3 and V_{BE}.

Fold-Back Current Limiting

The constant-current limiting technique just discussed restricts the load current to a *constant maximum value*. Another type of current limiting, known as *fold-back current limiting*, is used primarily in high-current regulators. When subjected to a short circuit or overload, the output current of the regulator with fold-back current limiting decreases to a level that is below the peak load current capability. In this manner, the power dissipation of the circuit is not exceeded.

Figure 7-6a shows a series regulator circuit with fold-back current limiting. The voltage across R_4

$$V_{R4} = V_{BE2} + V_{R5} \qquad \text{(Eq. 7-7)}$$

is due to the load current and must be sufficient to (1) overcome the base-emitter voltage needed to forward bias Q_2, as well as (2) overcome the voltage across R_5.

When the load current increases to a level that forces Q_2 to conduct, the load current is then limited at this point. The resulting decrease in the regulator's output voltage in turn forces a proportional decrease in voltage across R_5. Less current through R_4 is then required to keep Q_1 turned on. The output characteristic curve of Figure 7-6b shows that the load current decreases as the output voltage decreases. The short-circuit load current then decreases to a safe level that prevents the device from overheating.

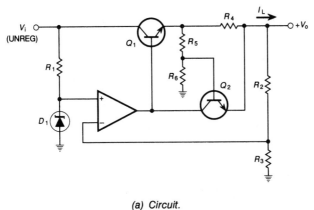

(a) Circuit.

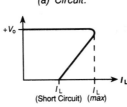

(b) Fold-back current characteristic.

Fig. 7-6. Series-pass regulator with fold-back current limiting.

THE SHUNT REGULATOR

The basic positive-voltage, low-power shunt regulator is shown in Figure 7-7. As with the series regulator, the output voltage of the op-amp is regulated at a voltage level equal to

$$V_o(REG) = \left(1 + \frac{R_2}{R_3}\right)V_{REF} \qquad \text{(Eq. 7-8)}$$

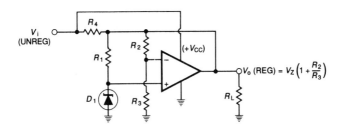

Fig. 7-7. Basic low-power shunt regulator.

The current-limiting resistor (R_1) for the zener diode is found from

$$R_1 = \frac{V_o - V_{REF}}{I_Z}$$ (Eq. 7-9)

If the output voltage tries to decrease in response to changes in either the input source voltage, load resistance, or temperature, R_2 and R_3 detect this level change and, in turn, feed it to the inverting input of the op-amp. Like the series regulator of Figure 7-3, the current through the voltage divider resistors (R_2 and R_3) can be assumed to be set at approximately 1 mA so as not to be loaded down by the input of the op-amp. In addition, the voltage across R_3 is the same as the zener voltage.

EXAMPLE 7-2

Using the shunt regulator circuit of Figure 7-7, determine the necessary component values for a regulated output of 12 V at a maximum load current of 10 mA using a 20-V unregulated source voltage and a 6.2-V zener diode biased at 5 mA.

For the current through D_1 to be 5 mA

$$R_1 = \frac{(12\ V - 6.2\ V)}{5\ mA}$$ (Eq. 7-9)

$$= 1.2\ k\Omega$$

For 1 mA flowing through R_2 and R_3

$$R_2 = \frac{12\ V - 6.2\ V}{1\ mA}$$ (Eq. 7-4)

$$= 5.8\ k\Omega \quad (use\ 5.6-k\Omega\ standard\ value)$$

$$R_3 = \frac{6.2\ V}{1\ mA}$$ (Eq. 7-5)

$$= 6.2\ k\Omega \quad (a\ standard\ value)$$

The current through R_4 is made up of the maximum load current, the current through the diode, and the current through R_2 and R_3

$$I = 10\ mA + 5\ mA + 1\ mA$$
$$= 16\ mA$$

so that

$$R_4 = \frac{20\ V - 12\ V}{16\ mA}$$
$$= 500\ \Omega \quad (use\ 510\text{-}\Omega\ standard\ value)$$

The final circuit is shown in Figure 7-8.

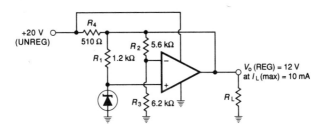

Fig. 7-8. Completed example circuit for a shunt regulator.

CURRENT REGULATORS

It is a simple matter to convert a voltage regulator to a *current regulator* or *constant-current source*, which delivers a fixed, constant current that is independent of load variations. A simple constant-current source is shown in Figure 7-9. This is really the same circuit as the voltage regulator of Figure 7-2. Here, the feedback resistor (R_2) of what is otherwise a noninverting amplifier is now the load. Since the input source voltage that forward biases the zener diode is

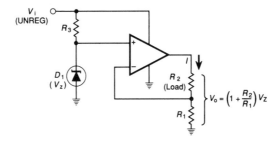

Fig. 7-9. Current regulator.

larger than the regulated input voltage (V_Z), the current through either R_1 or R_2 is essentially constant. Although the output voltage of the op-amp with respect to ground is always given as

$$V_o = \left(1 + \frac{R_2}{R_1}\right)V_Z \qquad \text{(Eq. 7-10)}$$

A problem with constant-current sources built around op-amps is that the load is floating. That is, neither terminal of the load is connected to ground. Use caution and be certain that a load will operate properly when connected to a floating ground.

Chapter 8

Waveform Generation

INTRODUCTION AND OBJECTIVES

When referring to signal sources, the *oscillator* is one that generates sine waves only. On the other hand, a *function generator* may not only provide sine waves, but also square waves, triangle waves, sawtooths, pulse trains, as well as any other type of periodic waveform. Both produce an output signal but require no external input signal.

At the completion of this chapter, you will be able to:

- Explain the difference between an oscillator and a function generator.
- Design and determine the output frequency of both a phase-shift and a Wien-bridge oscillator using op-amps, and how their outputs may be stabilized.
- Describe several methods for producing square waves, triangle waves, and sawtooth waveforms.

THE BASIC OSCILLATOR

An oscillator is a *sine-wave* signal source of known frequency. It produces an output signal, yet it requires no external input signal. The only external input connections to an oscillator are for the DC power source. The sine-wave output signal normally has both its frequency and amplitude determined by the circuit type, power supply voltage, and component values selected by the user.

The most common form of sine-wave oscillator consists of an amplifier with controlled *positive feedback*. As opposed to negative feedback in conventional amplifiers which tends to reduce the amplifier gain while providing circuit stability, positive feedback creates overall circuit instability by increasing amplifier gain.

RC OSCILLATORS

An RC oscillator uses one or more resistor-capacitor networks as its frequency selective feedback network. Because of the RC-component values as well as the frequency response of the op-amp used, its use is primarily for fixed frequency oscillators less than approximately 30 kHz where frequency stability is not much of a problem.

Phase-Shift Oscillator

Figure 8-1 shows the circuit of the *phase-shift oscillator*. The feedback network consists of three identical RC networks that collectively produce a 180° phase shift at

$$f_o = \frac{1}{2\pi RC\sqrt{6}}$$ (Eq. 8-1)

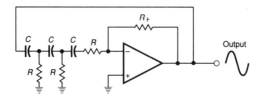

Fig. 8-1. Phase-shift oscillator.

At this frequency the RC-feedback network attenuates the amplifier's output by 1/29. For the loop gain to be unity (a condition necessary for oscillation), the inverting amplifier's closed-loop voltage gain must then be 29. Feedback resistor (R_F) is then made equal to 29 R. In practice however, the closed-loop gain is made slightly larger than 29 (i.e., 30 to 31) at the cost of a small amount of distortion. This is done to insure that the circuit will immediately start to oscillate when power is applied to the circuit.

The maximum output voltage and oscillation frequency for the oscillator circuit depends on several factors. The peak-to-peak output voltage will be approximately 2 volts less than the differential supply voltage ($V_{CC} + V_{EE}$). The maximum oscillation frequency is a function of the op-amp's gain-bandwidth product. As an example using a 741 op-amp (GBP = 1 MHz), the maximum oscillation frequency possible is then 1 MHz/29, or about 34.5 kHz.

As a general rule with most types of RC circuits, a standard capacitor value is first chosen. The corresponding value for the resistor is then calculated, settling on the nearest standard value with a very slight loss in accuracy. The phase-shift oscillator is primarily intended for fixed-frequency operation because the circuit requires that three equal components (either R or C) be simultaneously changed in a like manner to shift the oscillation frequency from one value to another.

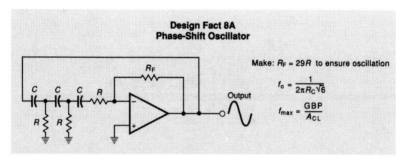

**Design Fact 8A
Phase-Shift Oscillator**

Make: $R_F = 29R$ to ensure oscillation

$$f_o = \frac{1}{2\pi R_C \sqrt{6}}$$

$$f_{max} = \frac{GBP}{A_{CL}}$$

EXAMPLE 8-1

Determine the suitable component values for a 1-kHz phase-shift oscillator.

Assuming 0.1-µF capacitors for C, then

$$R = \frac{1}{2\pi(1\ kHz)(0.1\ \mu F)(\sqrt{6})} \quad \text{(Eq. 8-1)}$$

$$= 650\ \Omega \quad (use\ 680\text{-}\Omega\ standard\ value)$$

so that

$$R_F = (29)(680\ \Omega)$$
$$= 19.7\ k\Omega \quad (use\ 20\text{-}k\Omega\ standard\ value)$$

Twin-T Oscillator

Another low-frequency RC oscillator is the *twin-T*, or *parallel-T oscillator*, as shown Figure 8-2. At resonance, the output frequency is given by

$$f_o = \frac{1}{2\pi RC} \qquad \text{(Eq. 8-2)}$$

In this circuit, the twin-T network provides negative feedback while incandescent lamp R_2 and resistor R_1 provide positive feedback. Because of inherent component mismatches, the variable resistor is used to adjust the twin-T network for optimum performance.

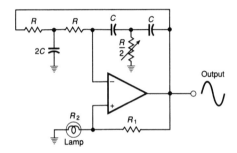

Fig. 8-2. Twin-T oscillator.

The bridge circuit requires that three component values (either R or C) be simultaneously changed to shift the oscillation frequency from one value to another. As this is awkward, the twin-T oscillator is primarily intended for fixed-frequency operation and is not easily tuned over a wide frequency range.

In comparison, the twin-T oscillator circuit is more stable than the phase-shift type.

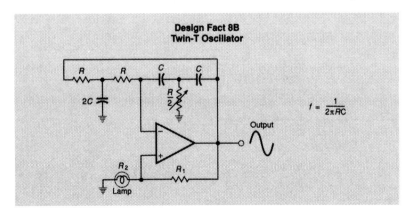

Wien-Bridge Oscillator

Figure 8-3 shows the *Wien-bridge oscillator*. Part of the Wien bridge is a dual RC network called a *lead-lag network*. Its name comes from the effect that the circuit has on the output signal. Below a certain frequency, the output signal is leading in phase with the input signal, while it lags the input signal above that frequency.

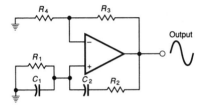

Fig. 8-3. Wien-bridge oscillator.

The Wien bridge itself as the frequency-selective RC-feedback network has a phase shift that varies from +90° to −90° as the frequency increases. The circuit oscillates at a frequency given by

$$f_o = \frac{1}{2\pi\sqrt{R_1 R_2 C_1 C_2}} \qquad \text{(Eq. 8-3)}$$

To simplify component selection, both frequency-determining resistors R_1 and R_2 are made equal, as are both frequency-determining capacitors C_1 and C_2. In this case, Equation 8-3 simplifies to

$$f_o = \frac{1}{2\pi R_1 C_1} \qquad \text{(Eq. 8-4)}$$

At this frequency the RC-feedback network attenuates the amplifier's output by ⅓. For the loop gain to be unity, the closed-loop voltage gain must then be 3. The lead-lag network offers a positive feedback path for an amplifier with a closed-loop gain of 3 and a loop gain of 1, thus creating the oscillator. The lead-lag network forms one side of a Wien bridge, and the gain-setting resistors R_3 and R_4 form the other side, so that for a closed-loop gain of 3

$$3 = 1 + \frac{R_3}{R_4}$$

or

$$R_3 = 2R_4 \qquad \text{(Eq. 8-5)}$$

Resistors R_3 and R_4 are in the negative feedback path and set the amplifier's closed-loop gain at 3. Resistor R_4 is often a device with a *positive* voltage-versus-resistance characteristic, such as an incandescent lamp in the 3.4-kHz oscillator circuit of Figure 8-4. This type of characteristic then tends to regulate the output amplitude. If the amplitude of the output voltage (V_o) starts to increase, R_4 increases to reduce the gain of the amplifier. If V_o decreases, R_4 decreases and the gain of the amplifier is increased. This action serves to maintain the output amplitude at a relatively constant level.

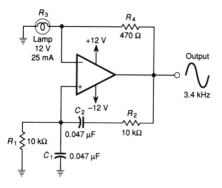

Fig. 8-4. Using an incandescent lamp having a positive voltage-vs.-resistance characteristic to regulate the output amplitude of a 3.4-kHz Wien-bridge oscillator.

Figure 8-5 shows another way of controlling the output amplitude of a Wien-bridge oscillator using back-to-back zener diodes placed across the negative feedback resistor R_4. When the oscillator is first turned on and the signal is small, the zener diodes do not conduct. Since potentiometer R_3 is initially adjusted to provide an amplifier closed-loop gain of 3, the signal increases in amplitude. When the signal peaks reach the zener breakdown voltage, their effective resistance is reduced, lowering the gain. This action then stabilizes the signal amplitude. The minimum output amplitude is controlled by the zener voltage.

If R_3 is adjusted for a smaller output, oscillations will cease. On the other hand, adjusting R_3 for a larger output increases the signal amplitude, but also increases the amount of distortion. Minimum distortion coincides with minimum amplitude.

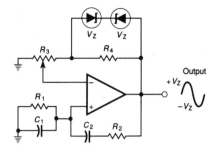

Fig. 8-5. Wien-bridge oscillator with back-to-back zener diodes used to limit output amplitude swings.

EXAMPLE 8-2

Calculate the output frequency of the Wien-bridge oscillator of Figure 8-3 with $C_1 = C_2 = 0.033$ μF, $R_1 = R_2 = 10$ kΩ, $R_3 = 1.5$ kΩ, and $R_4 = 3$ kΩ.

$$f_o = \frac{1}{2\pi(10\ k\Omega)(0.033\ \mu F)}$$ (Eq. 8-4)

$$= 483\ Hz$$

SQUARE-WAVE GENERATORS

Using any of the three RC oscillators discussed earlier in this chapter, a simple scheme used to generate square waves of the same frequency is the addition of a comparator to the oscillator's output. The comparator stage can be built using either a high slew rate op-amp like the LM318 (70 V/μs), or an integrated circuit comparator such as the LM710. If zener-diode limiting is not used, the peak output voltage levels for both the sine-wave and square-wave outputs are approximately $-V_{EE}$ and $+V_{CC}$.

An *astable multivibrator*, or *free-running multivibrator* is a circuit whose output has no stable state, or voltage level. Its output periodically *bounces* back and forth between two stable voltage levels. This then is precisely the action of a square wave so that an astable multivibrator is, in fact, another name for a square-wave generator.

Figure 8-6a shows an op-amp-based square-wave generator. When the op-amp output is in positive saturation, the noninverting input is held at $V_{SAT}/2$ by voltage divider resistors R_1 and R_2 acting as a positive feedback network. Meanwhile, timing capacitor C_1 is charging through R_3. When the capacitor voltage reaches $V_{-SAT}/2$, the inverting input becomes more positive than the noninverting input, and the op-amp's output switches to V_{-SAT} (Fig. 8-6b). This is similar to the action of an inverting comparator.

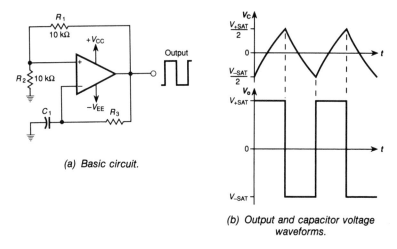

(a) Basic circuit.

(b) Output and capacitor voltage waveforms.

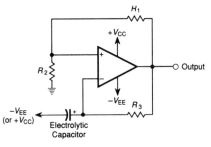

(c) Circuit connections when using an electrolytic capacitor.

Fig. 8-6. Op-amp square wave generator.

The noninverting input is now held at $V_{-SAT/2}$ until C_1 charges in the reverse direction to this level. When this occurs, the op-amp's output then switches back to positive saturation and the cycle repeats. This action produces a square wave that swings between the positive and negative saturation limits.

Besides the R_3-C_1 timing circuit, the output frequency, often called the free-running frequency, also depends on the positive feedback R_1-R_2 network, so that

$$f_o = \frac{1}{2R_3C_1 \ln\left[1 + (2R_2/R_1)\right]} \quad \text{(Eq. 8-6)}$$

If $R_1 = R_2$, then the free-running frequency can be written as

$$f_o = \frac{0.455}{R_3C_1} \quad \text{(Eq. 8-7)}$$

As an alternative approach, one can set $R_1 = 1.16\ R_2$, so that Equation 8-6 simplifies to

$$f_o = \frac{1}{2R_3C_1} \quad \text{(Eq. 8-8)}$$

As a general guide, R_1 and R_2 are selected to be in the range from 1 kΩ to 100 kΩ.

Several factors should be kept in mind for this type of circuit to function properly. First, since the op-amp functions essentially as a comparator, it should have both a high slew rate and should not be frequency compensated. This is because compensation reduces the frequency response and it is not necessary for stability. Finally, for low-frequency applications where C_1 is an electrolytic (polarized) type, the polarity of C_1 must be correct. As shown in Figure 8-6c, one end of C_1 must then be connected to either the negative or positive supply voltage, with the positive supply generally preferred.

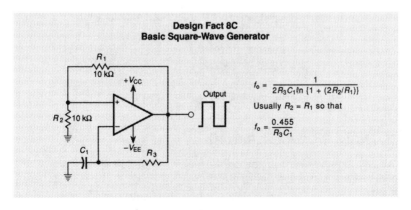

Design Fact 8C
Basic Square-Wave Generator

Figure 8-7 shows a similar square-wave generator circuit, but using a Norton-type op-amp. The free-running frequency is given by

$$f_o = \frac{0.707}{R_1 C_1} \qquad \text{(Eq. 8-9)}$$

if $R_3 = R_4$, and R_2 is made much larger than R_1. With the components shown, the free-running frequency is 216 Hz.

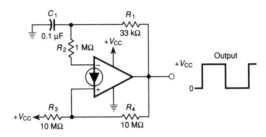

Fig. 8-7. Norton amplifier square-wave generator.

In Figure 8-8, the basic square-wave circuit of Figure 8-6 is modified to produce a nonsymmetrical pulse train. This is done by providing different charging rates for the timing capacitor during the positive and negative half cycles. The addition of the parallel D_1-R_4 path across R_3 allows the capacitor to charge at a much higher rate when the output is positive then when it is negative. As a result, the time that the output remains positive is shortened considerably.

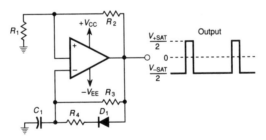

Fig. 8-8. Asymmetric pulse train generator.

TRIANGLE-WAVE GENERATORS

When a square wave is passed through an integrator, the output will be a triangular waveform having equal rise and fall times as well as a frequency equal to that of the astable's free-running frequency. Figure 8-9a shows the circuit for a triangle-wave generator using two op-amps. The first op-amp (A_1) is connected as an astable multivibrator and produces a square wave (V_o'). The second op-amp (A_2) is connected as a compensated integrator and converts the square wave to a triangle wave. Output errors due to A_2's input offset current are minimized by R_5, which typically is made equal to R_4.

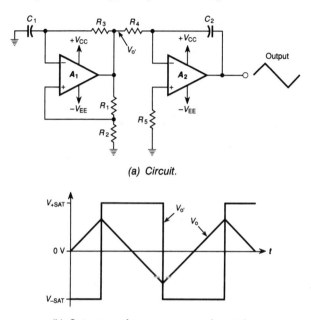

(a) Circuit.

(b) Output waveforms at op-amps A_1 and A_2.

Fig. 8-9. Triangle-wave generator.

Based on the level of the saturation voltage output of A_1, the peak-to-peak triangle output voltage of the integrator depends also on the frequency of the free-running multivibrator and the integrator's RC-time constant, so that

$$V_o(p-p) = \frac{V_{SAT}}{R_4 C_2 f_o}$$ (Eq. 8-10)

A modification of the circuit of Figure 8-9 is shown in Figure 8-10, which has a comparator (A_1) with hysteresis in conjunction with an integrator (A_2). When power is first applied to the circuit, the output of A_1 will be either at its positive or negative saturation voltage. Assuming that the output is negative, the output of the integrator will be a positive-going ramp. When the output of the integrator reaches the upper threshold voltage of A_1, as given by

$$V_{UT} = V_{-SAT} \frac{R_1}{R_2} \qquad \text{(Eq. 8-11)}$$

the output of the comparator will then switch to the positive saturation voltage, producing a negative-going ramp at the output of A_2. This continues until the output level of the integrator reaches A_1's lower threshold voltage, at which time the process repeats itself.

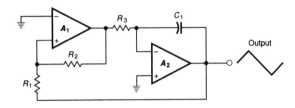

Fig. 8-10. Triangle-wave generator using a comparator with hysteresis and an integrator.

In terms of the circuit components, the output frequency of either the square or triangle waveforms is given by

$$f_o = \frac{R_2}{4R_1 R_3 C_1} \qquad \text{(Eq. 8-12)}$$

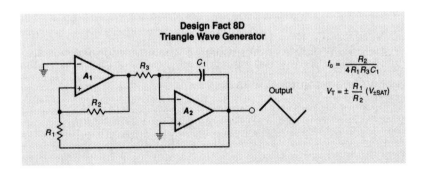

EXAMPLE 8-3

Using the triangle-wave generator of Figure 8-10, determine the necessary component values for an output frequency of 1 kHz and a peak-to-peak output voltage of 12 V if a +15-V supply is used.

Assume that the positive and negative saturation voltages are +13 V.

First, let's assume $R_1 = 1$ kΩ and $C_1 = 0.1$ μF as convenient starting points. From Equation 8-11, the lower (or upper) threshold level is one-half the desired peak-to-peak output, or +6 V. Solving for R_2 yields

$$R_2 = \left(\frac{13\ V}{6\ V}\right)(1\ k\Omega)$$

$$= 2.17\ k\Omega \quad (use\ 2.2\text{-}k\Omega\ standard\ value)$$

Finally, R_3 is found from Equation 8-12

$$R_3 = \frac{2.2\ k\Omega}{(4)(1\ k\Omega)(0.1\ \mu F)(1\ kHz)}$$

$$= 5.5\ k\Omega \quad (use\ 5.6\text{-}k\Omega\ standard\ value)$$

The final circuit is shown in Figure 8-11.

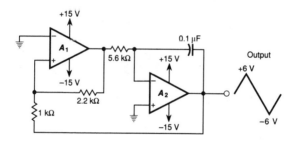

Fig. 8-11. Completed example circuit for a triangle-waveform generator.

Figure 8-12 shows a similar approach of the circuit of Figure 8-10, but with Norton amplifiers. Here, A_1 is an integrator which generates a linear ramp. Its output is fed to A_2, acting as a comparator so that its peak-to-peak output is set approximately equal to $V_{CC}/3$, while the output frequency is given by

$$f_o = \frac{1.5}{R_1 C_1} \quad \text{(Eq. 8-13)}$$

For the best symmetry, $R_2 = R_1/2$, while $R_4 = R_5 = 3R_3$.

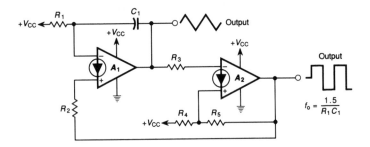

Fig. 8-12. Norton amplifier triangle-wave generator.

SAWTOOTH GENERATOR

The sawtooth waveform is a general case of the triangle wave where its rise and fall times are not equal. Figure 8-13 is a circuit that is a modification of Figure 8-10 that produces adjustable rise and fall times. Op-amp A_1 is connected as a noninverting comparator with hysteresis, while A_2 functions as an integrator.

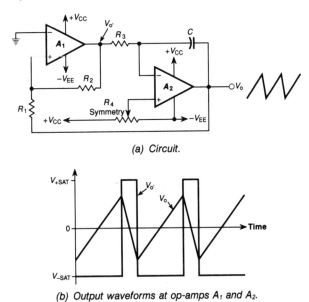

(a) Circuit.

(b) Output waveforms at op-amps A_1 and A_2.

Fig. 8-13. Sawtooth generator.

Chapter 9

Active Filters

INTRODUCTION AND OBJECTIVES

An active filter network built to use op-amps passes electric signals at certain frequencies or frequency ranges while preventing the passage of others. These may be used to intentionally remove unwanted noise or other frequency components that interfere with a desired signal, or they may be used to alter the shape of waveforms as in synthesizing musical sounds.

When compared with passive filters constructed only with inductors, resistors, and capacitors, active filters offer the following advantages:

- *No insertion loss.* Since the op-amp is capable of providing gain, the input signal will not be immediately attenuated as the filter passes those frequencies of interest.
- *Cost.* On the average, active filters will cost less than passive filters. This is because inductors are expensive and required values are not always available.
- *Tuning.* The frequency response of most active filters may be easily tuned, or adjusted over a wide frequency range without significantly changing the desired frequency response.
- *Isolation.* When using op-amps, active filters will have a high input impedance and a low output impedance, virtually guaranteeing almost no interaction between the filter and either its signal source or load.

On the other hand, there are a couple of disadvantages or limitations when using active filters.

1. *Frequency response.* The upper frequency response is limited by the frequency response of the transistor or op-amp. In general, the frequency response of transistors is greater than that of op-amps. Active filters are generally used for audio frequencies while most passive filters are used at RF.
2. *Power supply.* Unlike passive filters, active filters require some form of power supply for their operation.

At the completion of this chapter, you will be able to draw and calculate the required component values for a variety of RC-active filters using Sallen-Key, multiple-feedback, twin-T, and state-variable network designs.

FILTER RESPONSES

There are four basic filter frequency responses: low-pass, high-pass, bandpass, and notch. A low-pass filter allows incoming signals to be passed through with little or no attenuation up to its *cutoff frequency*.* Above this frequency, the filter rejects or greatly attenuates the level of the input signal. For the high-pass filter, the opposite is true.

How well the filter rejects input signals above or below the cutoff frequency is dependent on the *order* of the filter, which is a measure of how selective the filter is. The higher the order, the more selective the filter is. For Butterworth low-pass and high-pass filters, the filter rejects signals at a rate of an integer multiple of 6 dB/octave or 20 dB/decade. A 1st-order low-pass Butterworth filter has a rejection rate of 6 dB/octave or 20 dB/decade. A 2nd-order filter has a rejection rate that is twice as fast as 1st-order filter or 12 dB/octave (20 dB/decade), and so on.

The bandpass filter allows incoming signals to be passed through with little or no attenuation within a given median frequency range, or bandwidth, which is centered about a resonant frequency, called the *center frequency*. Above and below this range, the filter rejects or greatly attenuates the level of the input signal. How selective the filter is with respect to the center frequency is measured by its *quality factor* or Q. The Q is the ratio of the filter's center frequency to its bandwidth.

The notch filter operates in a manner that is the opposite of the bandpass filter and is primarily used to remove a signal at a single frequency, such as 60-Hz noise from an audio amplifier.

---

* The cutoff frequency is also referred to as either the *corner frequency*, *critical frequency*, *break frequency*, or *3-dB frequency*.

ACTIVE FILTER CIRCUITS

The basic active filter networks that are possible to realize various responses are

- *Sallen-Key.* Low-pass and high-pass.
- *Multiple-Feedback.* Low-pass, high-pass, and bandpass.
- *Twin-T.* Bandpass and notch.

All these are basic 2nd-order RC networks and are used as building blocks in the formation of higher-order filters. In addition, there are other specialized filter circuits, such as the state-variable filter, that are available.

Sallen-Key Networks

Figure 9-1a shows the basic 2nd-order *Sallen-Key filter* configuration, which is also called a *voltage-controlled voltage source* (VCVS). Depending upon the location of the specific frequency-determining resistors and capacitors for the four impedances (Z_1 through Z_4), it is possible to have a 2nd-order low-pass filter section (Figure 9-1b), and by interchanging the position of the RC components, create the equivalent high-pass network (Figure 9-1c).

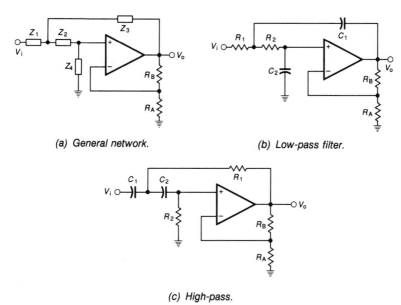

Fig. 9-1. 2nd-order Sallen-Key filter networks.

Based on the two resistors and capacitors, the cutoff frequency for both the low-pass and high-pass filters is given by

$$f_C = \frac{1}{2\pi\sqrt{R_1 R_2 C_1 C_2}} \qquad \text{(Eq. 9-1)}$$

For convenience, R_1 is made equal to R_2, while C_1 is made equal to C_2. This simplifies in what is called an *equal-component value VCVS filter*. Equation 9-1 then reduces to

$$f_C = \frac{1}{2\pi R_1 C_1} \qquad \text{(Eq. 9-2)}$$

A standard value for C_1 is chosen and the resulting value for R_1 is calculated for a given cutoff frequency from Equation 9-1. The *gain-setting resistors* R_A and R_B are chosen to give either a Bessel (maximum delay), Butterworth (maximally flat), or Chebyshev (equal ripple) response. Usually these two resistors are determined from the ratio R_B/R_A. For 2nd-order low-pass and high-pass filters, these ratios are summarized in Table 9-1. The required cutoff frequency for the filter section, as a factor of the desired cutoff frequency is also listed in Table 9-1.

Table 9-1. 2nd-Order VCVS Filter Parameters

Response	R_B/R_A	Low-pass	High-pass	Passband Gain
Bessel	0.268	$1.274f_C$	$0.785f_C$	1.27 (2.1 dB)
Butterworth	0.586	$1.000f_C$	$1.000f_C$	1.59 (4.0 dB)
1-dB Chebyshev	0.955	$0.863f_C$	$1.159f_C$	1.96 (5.8 dB)
2-dB Chebyshev	1.105	$0.852f_C$	$1.174f_C$	2.11 (6.5 dB)
3-dB Chebyshev	1.233	$0.841f_C$	$1.189f_C$	2.23 (7.0 dB)

Although Table 9-1 gives the necessary parameters for 2nd-order Bessel and Chebyshev filters, these 2nd-order responses are rarely used by themselves alone. The effectiveness of Bessel and Chebyshev filters are more pronounced in higher-order low-pass and high-pass filters.

EXAMPLE 9-1

Using the Sallen-Key circuit of Figure 9-1b, determine the necessary component values for a 2nd-order Butterworth low-pass filter having a cutoff frequency of 1.2 kHz. Assuming a standard value for C_1, such as 0.033 µF, then from Equation 9-2

$$R_1 = \frac{1}{2\pi(0.033 \ \mu F)(1.2 \ kHz)}$$

$$= 4{,}021 \ \Omega \quad (use \ 3.9\text{-}k\Omega \ standard \ value)$$

Assuming a standard value for R_A, such as 27 kΩ, then from Table 9-1, $R_B/R_A = 0.586$, so that R_B must be

$$R_B = (0.586)(27 \ k\Omega)$$

$$= 15.8 \ k\Omega \quad (use \ 15\text{-}k\Omega \ standard \ value)$$

In the passband for frequencies less than 1.2 kHz, the voltage gain from Table 9-1 is 1.59 (+4 dB). The final circuit is shown in Figure 9-2.

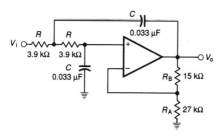

Fig. 9-2. Completed example circuit for a 2nd-order low-pass filter.

Higher-Order Filters

To obtain rolloff rates higher than can be achieved using the basic 2nd-order VCVS filter, higher-order filters are formed by cascading 1st- and 2nd-order filter sections (stages). The basic 1st-order low- and high-pass filter sections are shown in Figure 9-3a and 9-3b respectively. In both cases, the 1st-order active filter section is nothing more than a passive RC-filter network ahead of a voltage follower. Here, the passband voltage gain is unity and the rolloff is 6 dB/octave (or 20 dB/decade). In both cases, the cutoff frequency is the same as Equation 9-2.

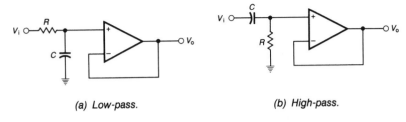

(a) Low-pass.　　　　　　　　(b) High-pass.

Fig. 9-3. *1st-order filter networks.*

The block diagrams of Figure 9-4 illustrate the schemes used to create higher-order low- and high-pass filters using 1st- and 2nd-order sections. Odd-order filters are achieved by cascading one 1st-order section with one or more 2nd-order sections. For example, a 5th-order high-pass filter is built by cascading one 1st-order high-pass filter section with two 2nd-order high-pass filter sections. Here, the order of the resultant filter ($N = 5$) is equal to the sum of the orders of the individual sections, so that $1 + 2 + 2 = 5$.

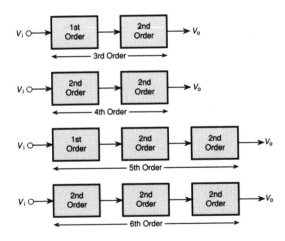

Fig. 9-4. **Creation of higher-order filters by cascading 1st- and 2nd-order filter sections.**

For even-ordered filters, only 2nd-order filter sections are used. Therefore, for an Nth-order low- or high-pass filter, only N/2 2nd-order sections are required. For example, a 6th-order high-pass filter requires 6/2, or three 2nd-order high-pass sections.

Table 9-2. Higher-Order Low-Pass Filter Parameters Using Cascaded Filter Sections

Order	1st Stage R_B/R_A	1st Stage $f_{C'}$*	2nd Stage R_B/R_A	2nd Stage $f_{C'}$*	3rd Stage R_B/R_A	3rd Stage $f_{C'}$*	Overall Passband dB Gain
\multicolumn{8}{c}{Bessel Response}							
3	–	1.328	0.553	1.454	–	–	3.8
4	0.084	1.436	0.759	1.610	–	–	5.6
5	–	1.557	0.225	1.613	0.909	1.819	7.4
6	0.041	0.298	0.364	0.722	1.023	0.975	9.2
\multicolumn{8}{c}{Butterworth Response}							
3	–	1	1.000	1	–	–	6.0
4	0.152	1	1.235	1	–	–	8.2
5	–	1	0.382	1	1.382	1	10.3
6	0.068	1	0.586	1	1.482	1	12.5
\multicolumn{8}{c}{1-dB Chebyshev Response}							
3	–	0.452	1.504	0.911	–	–	8.0
4	0.725	0.502	1.719	0.943	–	–	13.4
5	–	0.280	1.286	0.714	1.820	0.961	16.2
6	0.686	0.347	1.545	0.733	1.875	0.977	21.8
\multicolumn{8}{c}{2-dB Chebyshev Response}							
3	–	0.322	1.608	0.913	–	–	8.3
4	0.924	0.466	1.782	0.946	–	–	14.6
5	–	0.223	1.437	0.624	1.862	0.964	16.9
6	0.879	0.321	1.637	0.727	1.901	0.976	23.2
\multicolumn{8}{c}{3-dB Chebyshev Response}							
3	–	0.299	0.553	0.916	–	–	8.5
4	0.084	0.443	0.759	0.950	–	–	15.3
5	–	0.178	0.225	0.614	0.909	0.967	17.3
6	0.041	0.298	0.364	0.722	1.023	0.975	24.2

*The required factor for corresponding high-pass filter sections is obtained by taking the reciprocal.

The design of higher-order active filters is easily done by using tables which summarize the required design factors, as is done in Table 9-2. Here, the required cutoff frequencies for each filter stage are given for *low-pass filters* as a factor of the overall filter's cutoff frequency. From the table, the following characteristics should be noted:

1. To obtain the required frequencies for the corresponding high-pass filter, the reciprocal of the cutoff frequency factor is taken.
2. The first stage for all odd-order filters is always a 1st-order stage. The rest are 2nd-order stages.
3. For 1st-order filter stages, there is no value for R_B/R_A given as the op-amp is connected as a voltage follower.
4. For Butterworth filters, the cutoff frequency for all stages is identical and equal to the cutoff frequency of the overall higher-order filter.

EXAMPLE 9-2

Determine the necessary component values for a 3rd-order, 1-dB Chebyshev low-pass filter having a cutoff frequency of 941 Hz.

The cutoff frequency for the 1st-order first stage (Fig. 9-3) is

$$f_{C1'} = (0.452)(941\ Hz)$$
$$= 425\ Hz \quad \text{(refer to Table 9-2)}$$

Assuming a value for C_1, such that

$$C_1 = 0.022\ \mu F$$

then

$$R_1 = \frac{1}{2\pi(425\ Hz)(0.022\ \mu F)} \quad \text{(Eq. 9-2)}$$
$$= 17.03\ k\Omega \quad (\text{use } 18\text{-}k\Omega\ \text{standard value})$$

For the 2nd-order VCVS section, the cutoff frequency for the second stage is

$$f_{C2'} = (0.911)(941\ Hz)$$
$$= 857\ Hz \quad \text{(refer to Table 9-2)}$$

Again assuming a value for C_1, such that

$$C_1 = 0.01\ \mu F$$

then

$$R_1 = \frac{1}{2\pi(857\ Hz)(0.1\ \mu F)} \quad \text{(Eq. 9-2)}$$
$$= 1.86\ k\Omega \quad (use\ 1.8\text{-}k\Omega\ standard\ value)$$

Choosing a value for R_A, so that

$$R_A = 10\ k\Omega$$

then

$$R_B = (1.504)(10\ k\Omega)$$
$$= 15.04\ k\Omega \quad (use\ 15\text{-}k\Omega\ standard\ value) \quad \text{(refer to Table 9-2)}$$

The completed 3rd-order filter circuit composed of the 1st- and 2nd-order stages cascaded together is shown in Figure 9-5.

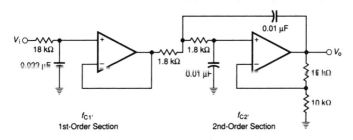

Fig. 9-5. Completed example circuit for a 3rd-order low-pass filter.

EXAMPLE 9-3

Determine the circuit and component values for a 4th-order Butterworth low-pass filter having a cutoff frequency of 1200 Hz.

For both 2nd-order VCVS sections, the cutoff frequency is the same as that for the overall 4th-order response (1200 Hz). Then assuming a value for C_1, such that $C_1 = 0.01\ \mu F$ for both sections

$$R_1 = \frac{1}{2\pi(1200\ Hz)(0.01\ \mu F)} \quad \text{(Eq. 9-2)}$$
$$= 13.3\ k\Omega \quad (use\ 13\text{-}k\Omega\ standard\ value)$$

For the first 2nd-order section

$$R_{B1}/R_{A1} = 0.152 \qquad \text{(refer to Table 9-2)}$$

Choosing a value for R_{A1} so that

$$R_{A1} = 10 \ k\Omega$$

then

$$\begin{aligned} R_{B1} &= (0.152)(10 \ k\Omega) \\ &= 1.52 \ k\Omega \quad (use \ 1.5\text{-}k\Omega \ standard \ value) \end{aligned}$$

For the second 2nd-order section

$$R_{B2}/R_{A2} = 1.235 \qquad \text{(refer to Table 9-2)}$$

Choosing a value for R_{A2} so that

$$R_{A2} = 12 \ k\Omega$$

then

$$\begin{aligned} R_{B2} &= (1.235)(12 \ k\Omega) \\ &= 14.8 \ k\Omega \quad (use \ 15\text{-}k\Omega \ standard \ value) \end{aligned}$$

The completed 4th-order filter circuit is shown in Figure 9-6.

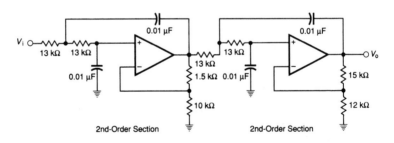

Fig. 9-6. Completed example circuit for a 4th-order low-pass filter.

Multiple-Feedback Networks

Besides the Sallen-Key network, another basic network that can be used for low- and high-pass, as well as bandpass filters is the *multi-*

ple-feedback filter network shown in Figure 9-7a. The cutoff frequency is determined from

$$f_C = \frac{1}{2\pi\sqrt{Z_2 Z_3 Z_4 Z_5}}$$ (Eq. 9-3)

Because there are five component values that must be chosen to simultaneously satisfy requirements of passband gain, cutoff frequency, and the choice of either a Bessel, Butterworth, or Chebyshev response, there are no easy simplifications in the relationships between the five components that can be made. Whenever a simplification (such as equal capacitor values) is tried, some of the remaining requirements are then hard to achieve.

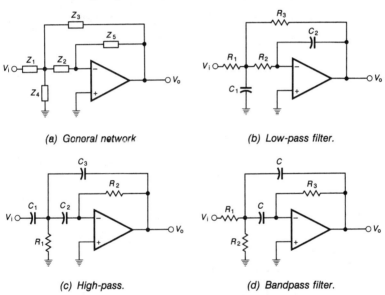

(a) General network

(b) Low-pass filter.

(c) High-pass.

(d) Bandpass filter.

Fig. 9-7. 2nd-order multiple-feedback filter networks.

For these reasons, the multiple-feedback filter network is best suited for bandpass filters (Fig. 9-7d) for Q's up to approximately 15. Here, component selection can be simplified, such that both capacitors can be chosen to have the same value. When this is the case, the center frequency is given by

$$f_o = \frac{1}{2\pi C} \sqrt{\frac{R_1 + R_2}{R_1 R_2 R_3}}$$ (Eq. 9-4)

155

For the given values of capacitance (C), Q, center frequency (f_o), and passband voltage gain (A_o), the values of the three resistors are found from

$$R_1 = \frac{Q}{2\pi f_o C A_o} \qquad \text{(Eq. 9-5)}$$

$$R_2 = \frac{Q}{2\pi f_o C (2Q^2 - A_o)} \qquad \text{(Eq. 9-6)}$$

$$R_3 = \frac{Q}{\pi f_o C} \qquad \text{(Eq. 9-7)}$$

Because of the interaction between Q and the center frequency gain in Equation 9-6, the following must hold

$$Q > \sqrt{\frac{A_o}{2}} \qquad \text{(Eq. 9-8)}$$

while the center frequency gain is determined from

$$A_o = \frac{R_3}{2R_1} \qquad \text{(Eq. 9-9)}$$

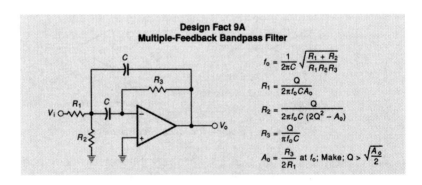

Design Fact 9A
Multiple-Feedback Bandpass Filter

$f_o = \frac{1}{2\pi C} \sqrt{\frac{R_1 + R_2}{R_1 R_2 R_3}}$

$R_1 = \frac{Q}{2\pi f_o C A_o}$

$R_2 = \frac{Q}{2\pi f_o C (2Q^2 - A_o)}$

$R_3 = \frac{Q}{\pi f_o C}$

$A_o = \frac{R_3}{2R_1}$ at f_o; Make; $Q > \sqrt{\frac{A_o}{2}}$

EXAMPLE 9-4

Determine the required component values for a multiple-feedback bandpass filter having a center frequency of 750 Hz, Q of 4.2, and a center frequency gain of 1.3.

Equation 9-8 must first be used to check whether or not this circuit can have a Q of 4.2 with a passband gain of 1.3

$$Q > \sqrt{\frac{1.32}{2}}$$

$$4.2 > 0.81$$

which is sufficient.

Choosing a standard capacitor value of 0.01 µF, the three resistors are determined

$$R_1 = \frac{4.2}{2\pi(750\ Hz)(0.01\ \mu F)(1.3)} \quad \text{(Eq. 9-5)}$$

$$= 68.6\ k\Omega \quad (use\ 68\text{-}k\Omega\ standard\ value)$$

$$R_2 = \frac{4.2}{2\pi(750\ Hz)(0.01\ \mu F)\big((2)(4.2)^2 - 1.32\big)} \quad \text{(Eq. 9-6)}$$

$$= 2.6\ k\Omega \quad (use\ 2.7\text{-}k\Omega\ standard\ value)$$

$$R_3 = (2)(68.6\ k\Omega)(1.3)$$
$$= 178\ k\Omega \quad (use\ 180\text{-}k\Omega\ standard\ value) \quad \text{(Eq. 9-9)}$$

The completed circuit is shown in Figure 9-8.

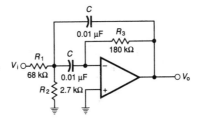

Fig. 9-8. Completed example circuit for a multiple-feedback bandpass filter.

From classical filter theory, one way of forming a notch filter is to subtract the output signal of a bandpass filter from its input signal. The basic multiple-feedback bandpass filter of Figure 9-7d can be combined with a 2-input summing amplifier to form a notch filter circuit, as shown in Figure 9-9. Here, the design of the circuit is identical to the bandpass circuit for Q and f_o.

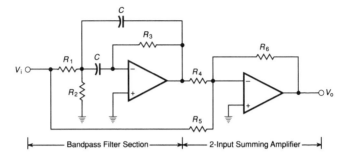

Fig. 9-9. Notch filter formed by summing the output of a bandpass filter network with the filter's input signal.

At the bandpass filter section's center frequency, the output signal is 180° out of phase with the input. The summing amplifier then subtracts the inverted bandpass output signal from the input signal of the filter. To produce the deepest null possible at the filter's center frequency, we require that

$$\frac{R_5}{R_4} = \frac{R_3}{2R_1}$$ (Eq. 9-10)

and

$$R_5 = R_6$$ (Eq. 9-11)

Notice that the right-hand term of Equation 9-10 ($R_3/2R_1$) is simply the same as the center frequency passband gain of the bandpass filter section. The overall passband gain of the resulting notch filter circuit equals 1.

Twin-T Filters

The *twin-T*, or *parallel-T network* is a popular configuration that can be used in the formation of both bandpass and notch filters (Fig. 9-10). For these circuits to work properly, the two capacitors in the *upper T* path should each be exactly twice the value of the capacitor in the *lower T* network. For the resistors, the opposite is true. Based on R and C, the center frequency is determined by

$$f_o = \frac{1}{2\pi RC} \qquad \text{(Eq. 9-12)}$$

At the center frequency, the voltage gain is limited by the ratio R_B/R_A, which is the closed-loop inverting gain of the op-amp.

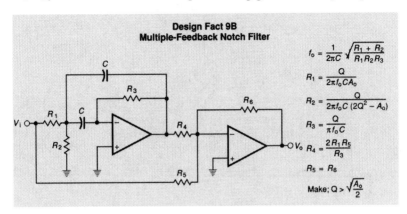

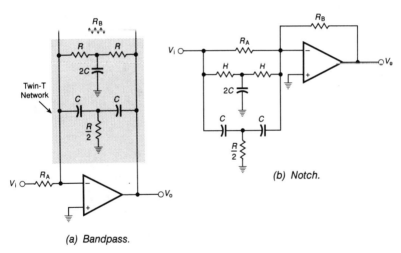

Fig. 9-10. Twin-T filters.

The twin-T notch filter of Figure 9-10b is similar to the bandpass version except that the twin-T network is placed in the input path instead of the feedback path. At frequencies other than the filter's center frequency, the closed-loop voltage gain is limited by the ratio R_B/R_A.

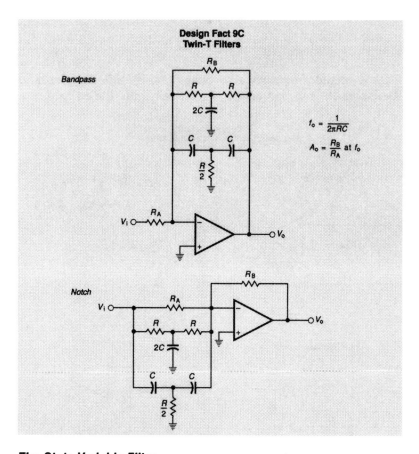

The State-Variable Filter

The filter shown in Figure 9-11 is called a *state-variable,* or *universal filter.* It simultaneously provides separate 2nd-order low- and high-pass outputs in addition to a high-Q bandpass filter output. Such a filter is composed of a difference amplifier (A_1) and two identical integrators (A_2 and A_3).

For this circuit, the center frequency of the bandpass response and cutoff frequency of either the low- or high-pass output are the same, so that

$$f_o = \frac{1}{2\pi RC} \qquad \text{(Eq. 9-13)}$$

The filter's Q is set solely by resistors R_A and R_B, so that

$$R_A = (3Q - 1)R_B \qquad \text{(Eq. 9-14)}$$

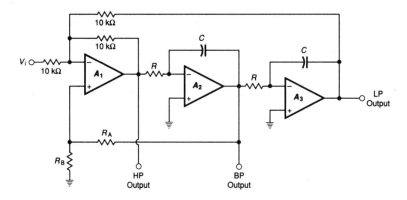

Fig. 9-11. State-variable filter having simultaneous low-pass, high-pass, and bandpass outputs.

With such a filter, Q's up to several hundred can be achieved with stability. For the low- and high-pass outputs, the passband gain is unity. However, for the bandpass response, the center frequency gain equals the filter's Q, so that high Q results in high gain and care must be taken to insure that the op-amp's gain-bandwidth product specification is observed.

As this type of filter gives a 2nd-order response for both the low- and high-pass outputs, it may not be possible to obtain optimum performance with all three outputs simultaneously. For a 2nd-order Butterworth low- or high-pass response, $Q = 0.707$, such that

$$R_A = 1.12\, R_B \qquad \text{(Eq. 9-15)}$$

Consequently, one designs either for a Butterworth low-pass/high-pass response ($Q = 0.707$) or a high-Q bandpass response.

EXAMPLE 9-5

Determine the necessary component values for the state-variable filter whose bandpass response is to have a center frequency of 800 Hz and a Q of 40.

Choosing a 0.01-µF standard capacitor value for C, then

$$R = \frac{1}{2\pi (0.01\ \mu F)(800\ Hz)} \qquad \text{(Eq. 9-13)}$$

$$= 19.9\ k\Omega \qquad (\textit{use 20-k}\Omega\ \textit{standard value})$$

Choosing a 1.8 kΩ standard resistor value for R_B, then

$$R_A = [(3)(40) - 1](1.8\ k\Omega)$$
$$= 214.2\ k\Omega \quad (use\ 220\text{-}k\Omega\ standard\ value)$$

The completed circuit is shown in Figure 9-12.

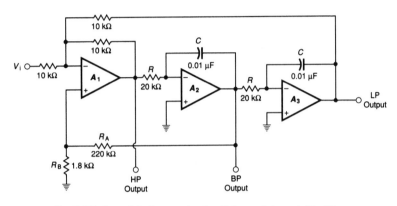

Fig. 9-12. Completed example circuit for a state-variable filter.

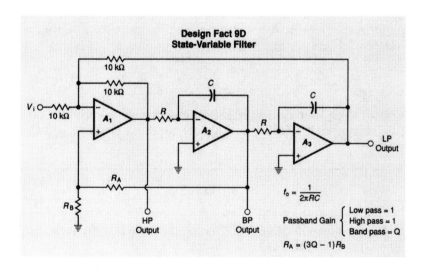

One very nice feature of the basic state-variable filter of Figure 9-11 is a notch filter can be formed by simultaneously adding its low- and high-pass outputs with equal weighting. As shown in Figure 9-13, this circuit results in a state-variable notch filter and requires only an additional op-amp and three equal-valued resistors which are typically 10 kΩ for convenience.

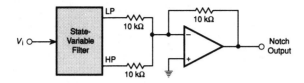

Fig. 9-13. State-variable notch filter formed by summing low-pass and highpass outputs.

The passband voltage gain is unity, and the notch frequency is the same as the basic state-variable filter's cutoff frequency (or center frequency), while the filter Q is the same for the bandpass filter response.

Chapter 10

Experiments

INTRODUCTION

This chapter allows you to perform 14 simple experiments that demonstrate principles, concepts, and applications for many basic circuit configurations using op-amps.

PERFORMING THE EXPERIMENTS

A laboratory experiment is a powerful learning tool in the educational process and is a double-edged sword. In order to receive the benefits it can provide, then you must follow several rules so that the experiment will be successful. The following sections discuss these rules and describe the format of how each experiment is presented.

Breadboarding

The breadboard is designed to accommodate the experiments that you will perform. The various op-amps, transistors, diodes, resistors, capacitors, other components, as well as power and signal connections all tie directly to the breadboard. Figure 10-1 shows the top view of a solderless breadboarding socket, which is manufactured by several companies.

Breadboarding is an art that cannot be learned in a few minutes, but takes practice and experience to develop an efficient technique. Just as an artist plans his creation, making sure that the picture will fit on the canvas in the proper proportions without crowding, the same is true with breadboarding electronic circuits.

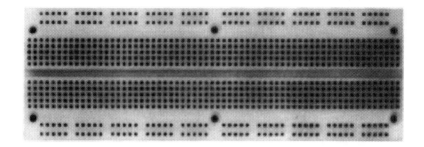

Fig. 10-1. Solderless breadboard socket.

When breadboarding, keep the following rules in mind:

1. Only No. 22, 24, or 26 insulated wire should be used, and it must be *solid*, not stranded.
2. Never insert too large a wire or component lead into a breadboarding terminal.
3. Never insert a bent wire. Straighten out the bent end with a pair of pliers before insertion.
4. Try to maintain an orderly arrangement of components and wires, keeping all connections as short as possible.

    Generally, the circuit is arranged on the breadboard in the same way that it appears on a schematic. This is useful when trying to locate possible wiring errors.

Setting Up the Experiments

Throughout this laboratory workbook, you will have the opportunity to breadboard a variety of circuits. Before you set up any experiment, you should do the following:

1. Plan your experiment beforehand. Know what types of results you are expected to observe.
2. Disconnect, or turn off *all* power and external signal sources from the breadboard.
3. Clear the breadboard of all wires and components from previous experiments, unless instructed otherwise.
4. Check the wired-up circuit against the schematic to make sure that it is correct.

5. Unless otherwise instructed, never make component or wiring changes on or to the breadboard with the power or external signal connections to the breadboard. This rule reduces the possibility of accidentally destroying electronic components and equipment.
6. When finished, make sure that you disconnect all power and signal sources before you clear the breadboard of wires and components.

Format for the Experiments

The instructions for each experiment are presented in the following format:

1. *Purpose.* The material under this heading states a brief purpose for performing the experiment. You should have this intended purpose in mind as you conduct the experiment.
2. *Schematic of Circuit.* The schematic of the completed circuit that you will construct for the experiments is given. You should analyze this diagram in an effort to obtain an understanding of the circuit *before* you proceed further. When used, oscilloscope and meter connections to the circuit are shown with bolder lines so that they will not be confused with normal circuit connections.
3. *Formulas.* Under this heading is a summary of the equations that apply to the design and/or operation of the circuit. These formulas are presented so that you can calculate results or compare measured results with theory.
4. *Procedure.* A series of numbered, sequential steps describe the detailed instructions for performing portions of the experiment. When appropriate, the initial settings for the oscilloscope are given as an aid. These settings, however, must not be considered constant and can be changed at any time to suit your preferences. For best results, a digital multimeter should be used. Any numerical calculations are performed easily on many of the pocket-type calculators.

Most experiments will ask you to compare either a measured value or a calculated value based on other known measured values with an expected, or true value. The expected value is one you would expect to obtain if you had not done the experiment. Generally this would be determined from theory.

EXPERIMENT 1—VOLTAGE FOLLOWER, NONINVERTING, AND INVERTING AMPLIFIERS

Purpose

The purpose of this experiment is to demonstrate and compare the operation of the voltage follower, noninverting, and inverting amplifier circuits using the 741 op-amp. In all cases in this experiment, the op-amp can be treated as an ideal device.

Schematic of Circuit

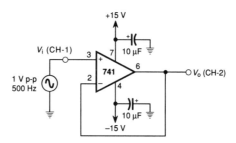

Fig. 10-2. Voltage follower for Experiment 1.

Formulas

Noninverting amplifier closed-loop voltage gain

$$A_{CL} = 1 + \frac{R_2}{R_1} \qquad \text{(Eq. 10-1)}$$

Inverting amplifier closed-loop voltage gain

$$A_{CL} = -\frac{R_2}{R_1} \qquad \text{(Eq. 10-2)}$$

Step 1

Wire the voltage follower circuit shown in Figure 10-2. For this and the remainder of the experiments using op-amps, the pair of 10-μF capacitors will be used to bypass or decouple the power supply leads. Make sure that the polarity of these capacitors are wired correctly. Set your oscilloscope for the following approximate settings:

Channels 1 and 2: 0.5 V division, AC coupling
Time Base: 1 ms/division

Step 2

Apply the signal generator and power supply voltages to the breadboard and adjust the peak-to-peak sine-wave input voltage (V_i) at 1 V and its frequency to 500 Hz. Position the input signal above the output signal (V_o) on the oscilloscope's display. Are there any differences between the two signals?

You should notice that the output signal appears exactly like the input. Both peak-to-peak voltages and waveshapes are the same. The input signal is fed to the op-amp's noninverting input so that the instantaneous output signal will have the same polarity as the input.

Step 3

Now, arbitrarily increase or decrease both the amplitude and frequency of the input signal. You should find that the output exactly follows the input. Also, change the input signal from a sine wave to a square or triangle waveform. You should notice that, regardless of the input waveform's amplitude or shape, the output signal still looks the same as the input. The output signal then exactly follows any changes in the input signal.

Step 4

Turn off and disconnect both the signal generator and the power supply from the breadboard and wire the noninverting amplifier circuit of Figure 10-3. Also, set your oscilloscope to the same approximate settings given in Step 1.

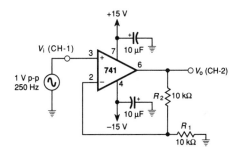

Fig. 10-3. Noninverting amplifier for Experiment 1.

Step 5

Apply the power supply and signal generator voltages to the breadboard and adjust the peak-to-peak input voltage at 1 V and its frequency to 1 kHz. Again position the input voltage above the output voltage on the oscilloscope's display. What is the difference between the two signals?

You should observe that the only difference is that the output signal is larger than the input signal. Both signals are said to be *in phase,* since the output signal goes positive exactly when the input does.

Step 6

To measure the amplifier's voltage gain, you can either measure the peak-to-peak output voltage and divide it by the peak-to-peak input signal, or you can measure both the input and output RMS voltages with an AC voltmeter, since the AC signal is a sine wave. Using either approach, calculate the experimental voltage gain, compare it to the expected value (Eq. 10-1), and record your results in Table 10-1.

Table 10-1.

R_2	Measured V_o	Measured Gain	Expected Gain
10 kΩ			
27 kΩ			
47 kΩ			
100 kΩ			
5.6 kΩ			
1 kΩ			

Step 7

Keeping the peak-to-peak input signal at 1 V, change resistor R_2 to the five remaining values listed in Table 10-1. For each value record your results in the table. Each time, turn off the power supplies and signal generator connections to the breadboard before you change the resistor. Do your values agree with the noninverting amplifier voltage gain equation?

You should observe that, from the values of the table, the voltage gain of a noninverting amplifier can never be less than 1, or *exactly* equal to 1. It will always be *greater* than 1.

Step 8

Turn off and disconnect the power and signal generator from the breadboard. Now wire the inverting amplifier circuit shown in the schematic of Figure 10-4. Apply power to the breadboard and adjust the input voltage at 1 V peak-to-peak and its frequency to 500 Hz. As before, position the input voltage above the output voltage on the oscilloscope display. What are the differences between the two signals?

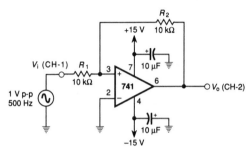

Fig. 10-4. Inverting amplifier for Experiment 1.

You should observe that the output signal is of opposite form, or is *inverted*, compared with the input signal. The output voltage is then said to be inverted, or 180° out of phase with the input, since the positive peak of the output signal occurs when the input's peak is negative.

Step 9

As in Step 6, measure the amplifier's voltage gain by either measuring the peak-to-peak output voltage or by measuring the input and output RMS voltages with an AC voltmeter. Compare the measured voltage gain with the expected value (Eq. 10-2), recording your values in Table 10-2.

You should find that the peak-to-peak output voltage should be nearly the same as the input, so that the voltage gain is −1. The minus sign indicates that the output is inverted with respect to the input, which represents a 180° phase shift.

Step 10

Keeping the input signal at 1 V peak-to-peak, change resistor R_2 to the remaining five values listed in Table 10-2, recording your values each time as in Step 9. Turn off the power supplies and signal generator *before* you change the resistor. You should notice that the voltage gain of an inverting amplifier can be less than 1, equal to 1, or greater than 1.

Table 10-2.

R_2	Measured V_o	Measured Gain	Expected Gain
10 kΩ			
27 kΩ			
47 kΩ			
100 kΩ			
5.6 kΩ			
1 kΩ			

EXPERIMENT 2—THE SUMMING AMPLIFIER/AVERAGER

Purpose

The purpose of this experiment is to demonstrate the operation of a 3-input summing amplifier and a variation which allows the determination of the algebraic average voltage value of 3 inputs.

Schematic of Circuit

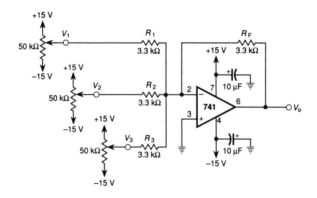

Fig. 10-5. Summing amplifier for Experiment 2.

Formulas

Three-input summing amplifier output voltage

$$V_o = -\left(\frac{R_F}{R_1}\right)V_1 - \left(\frac{R_F}{R_2}\right)V_2 - \left(\frac{R_F}{R_3}\right)V_3 \quad \text{(Eq. 10-3)}$$

$$V_o = -\frac{R_F}{R_1}(V_1 + V_2 + V_3) \quad (\text{if } R_1 = R_2 = R_3) \text{ (Eq. 10-4)}$$

Step 1

Wire the summing amplifier circuit shown in the schematic of Figure 10-5 and apply power to the breadboard. Using a DC voltmeter, adjust the three potentiometers so that $V_1 = -4$ V, $V_2 = +2$ V, and $V_3 = +5$ V. Now measure the DC output voltage and record its value in Table 10-3. How does it compare with the expected value (Eq. 10-3)?

Table 10-3.

V_1	V_2	V_3	Measured V_o	Expected V_o
−4 V	+2 V	+5 V		
+1 V	+2 V	−2 V		
+3 V	−3 V	−4 V		
+1 V	+1 V	−1 V		

Step 2

Now adjust all three potentiometers to obtain the remaining three DC voltages for V_1, V_2, and V_3 listed in Table 10-3. In each case, measure the corresponding DC output voltage of the summing amplifier and compare it with the expected value.

Step 3

Turn off and disconnect the power from the breadboard and wire the summing amplifier circuit so that all three input resistors (R_1, R_2, and R_3) are 3.3 kΩ, while the feedback resistor (R_F) is now 1 kΩ. Adjust all three potentiometers to obtain the DC voltages for the values of V_1, V_2, and V_3 listed in Table 10-4. Measure the corresponding DC output voltage of the summing amplifier and compare it with the expected value (Eq. 10-4). Record all values in the table.

Table 10-4.

V_1	V_2	V_3	Measured V_o	Expected V_o
−4 V	+2 V	+5 V		
+1 V	+2 V	−6 V		
−3 V	−3 V	−6 V		
+1 V	+2 V	+3 V		

In each case, you should measure output voltages that are equal to the algebraic average of the three input voltages, except that the output polarity is inverted, or is opposite of what you might expect the average to be. A summing amplifier having N equal input resistors can be changed to take the average of its inputs by setting the value of the feedback resistor to $1/N$ times the value any one of the input resistors.

EXPERIMENT 3—THE DIFFERENCE AMPLIFIER AND COMMON-MODE REJECTION

Purpose

The purpose of this experiment is to (1) measure the common-mode rejection of the 741 op-amp, (2) demonstrate the operation of a difference amplifier, and (3) demonstrate how to maximize the difference amplifier's common-mode rejection.

Schematic of Circuit

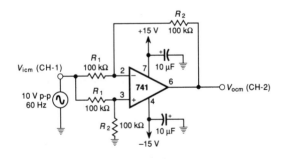

Fig. 10-6. Difference amplifier for Experiment 3.

Formulas

Amplifier differential voltage gain

$$A_D = \frac{R_2}{R_1} \qquad \text{(Eq. 10-5)}$$

Common-mode voltage gain

$$A_{cm} = \frac{V_{ocm}}{V_{icm}} \qquad \text{(Eq. 10-6)}$$

Common-mode rejection ratio (CMRR)

$$CMRR = \frac{A_D}{A_{cm}} \qquad \text{(Eq. 10-7)}$$

dB common-mode rejection (CMR)

$$CMR\ (dB) = 20\ \log_{10}(CMRR) \qquad \text{(Eq. 10-8)}$$

Differential amplifier output voltage

$$V_o = (V_2 - V_1)A_D + \frac{A_D}{CMRR} V_{icm} \qquad (actual) \qquad \text{(Eq. 10-9)}$$

$$V_o = (V_2 - V_1)A_D \qquad (ideal) \qquad \text{(Eq. 10-10)}$$

Step 1

Wire the circuit shown in the schematic of Figure 10-6 and set your oscilloscope for the following approximate settings:

Channel 1: 2 V/division, AC coupling
Channel 2: 0.2 V/division, AC coupling
Time Base: 5 ms/division, chopped triggering.

Step 2

Apply power to the breadboard and adjust the input voltage, called the *common-mode input voltage* V_{icm} to 10 V peak-to-peak at a frequency of approximately 60 Hz. With an AC voltmeter, measure the RMS common-mode input voltage V_{ocm} and record this value in Table 10-5.

Table 10-5.

Measured RMS common-mode input voltage, V_{icm}	V
Measured RMS common-mode output voltage, V_{ocm}	V
Calculated common-mode gain, A_{cm}	
Differential voltage gain, A_D	1000
Calculated common-mode rejection, CMR	dB

Step 3

Now measure the corresponding RMS common-mode output voltage, V_{ocm}, and record this result in Table 10-5. You may have to increase the sensitivity of the multimeter to make this reading as accurate as possible. From the measured common-mode input and output voltages, calculate the common-mode voltage gain, A_{cm} (Eq. 10-6), and record this value in the table.

This circuit is called a *difference*, or *differential amplifier*. First calculate the differential voltage gain, then use it to calculate the common-mode rejection (in decibels) for your particular 741 op-amp and record this value in Table 10-5. Most manufacturers of the 741 op-amp cite a minimum CMR of 70 dB, with typical value of 90 dB.

Step 4

In some cases, the CMR of a given difference amplifier circuit can be substantially improved by trimming one or more resistors of the circuit so that both resistance ratios (R_2/R_1) are the same. To do this, turn off and disconnect the signal and power leads from the circuit. Then replace the 100-kΩ resistor connected to the noninverting input with a series connection consisting of a 47-kΩ resistor and a 100-kΩ potentiometer.

Step 5

Apply power to the breadboard and adjust the common-mode input voltage at 10 V peak-to-peak and its frequency to 60 Hz.

Step 6

Using the oscilloscope to observe the amplifier's common-mode output signal, adjust the 100-kΩ potentiometer for a minimum peak-to-peak output voltage.

Step 7

Repeat Steps 2 and 3 using a differential gain of 1000 and record these values in Table 10-6. Do you see any improvement in the CMR?

Table 10-6.

Measured RMS common-mode input voltage, V_{icm}	V
Measured RMS common-mode output voltage, V_{ocm}	V
Calculated common-mode gain, A_{cm}	
Differential voltage gain, A_D	1000
Calculated common-mode rejection, CMR	dB

Step 8

Turn off and disconnect the power and signal generator from the breadboard. Now wire the differential amplifier circuit shown in Figure 10-7.

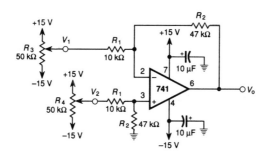

Fig. 10-7. *Another difference amplifier for Experiment 3.*

Step 9

Apply power to the breadboard. Using a DC voltmeter, adjust potentiometer R_3 so that V_1 is +4 V. Then adjust potentiometer R_4 so that V_2 is +6 V. Now measure the DC output voltage and record its value in Table 10-7. How does it compare with the expected value (Eq. 10-10)?

Since the differential gain is 4.7, the DC output voltage should be 4.7 times larger than the 2-V differential input voltage ($V_2 - V_1$), or +9.4 V.

177

Table 10-7.

V_1	V_2	Measured V_o	Expected V_o
+4 V	+6 V		
+1 V	−5 V		
−5 V	−2 V		
−1 V	−4 V		

Step 10

Now adjust both potentiometers to obtain the remaining three DC voltages for V_1 and V_2 listed in Table 10-7. In each case, measure the corresponding output voltage of the difference amplifier and compare it with the expected value. Record all values in the table.

EXPERIMENT 4—MEASUREMENT OF OPERATIONAL AMPLIFIER PARAMETERS

Purpose

The purpose of this experiment is to measure several parameters which can seriously effect the performance of a given op-amp. These are the input offset voltage, input bias current, input offset current, and the slew rate of a 741 op-amp.

Schematic of Circuit

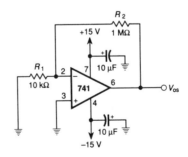

Fig. 10-8. Circuit for the measurement of input offset voltage.

Formulas

Input offset voltage

$$V_{oi} = \frac{V_{os}}{A_{CL}} \qquad \text{(Eq. 10-11)}$$

Bias currents

$$I_{B1} = \frac{V_1}{R_1} \qquad \text{(Eq. 10-12a)}$$

$$I_{B2} = \frac{V_2}{R_2} \qquad \text{(Eq. 10-12b)}$$

Input bias current

$$I_B = \frac{I_{B1} + I_{B2}}{2} \qquad \text{(Eq. 10-13)}$$

Input offset current

$$I_{os} = |I_{B2} - I_{B2}| \qquad \text{(Eq. 10-14)}$$

Step 1
Wire the circuit shown in the schematic of Figure 10-8 to measure the input offset voltage.

Step 2
Apply power to the breadboard. With your DC voltmeter, measure the DC output offset voltage (V_{os}) at pin 6 of the 741 op-amp and record this value in Table 10-8.

Table 10-8.

Measured DC output offset voltage (V_{os})	mV
Closed-loop voltage gain A_{CL}	
Calculated input offset voltage (V_{oi})	mV

Step 3

Calculate the input offset voltage (Eq. 10-11) for your particular op-amp and record this value in Table 10-8. For the 741 op-amp, the input offset voltage is typically 2 mV, with a maximum of 6 mV.

Step 4

Disconnect the power from the breadboard. Wire the circuit shown in the schematic of Figure 10-9 to measure the input bias current and input offset current

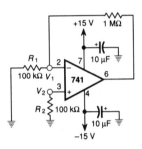

Fig. 10-9. Circuit for the measurement of input bias current and input offset current.

Step 5

Apply power to the breadboard. With a DC voltmeter, measure the DC voltage (V_1) across resistor R_1. Then measure the DC voltage (V_2) across resistor R_2 and record both values in Table 10-9.

Table 10-9.

Measured voltage across R_1 (V_1)	mV
Measured voltage across R_2 (V_2)	mV
Calculated input bias current (I_{B1})	nA
Calculated input bias current (I_{B2})	nA
Calculated average input bias current (I_B)	nA
Calculated input offset current (I_{os})	nA

Step 6

Calculate the bias current (Eq. 10-12a and 10-12b) flowing into each input of your particular op-amp and record your results in Table 10-9. Then calculate the input bias current for the op-amp (Eq. 10-13), which is the *average* of the two bias currents, and record this value in the table.

Step 7

Calculate the input offset current (Eq. 10-14) and record this value in Table 10-9.

For the 741 op-amp the input bias current is typically 80 nA, with a maximum of 500 nA. The input offset current is typically 20 nA, with a maximum of 200 nA.

Step 8

Turn off and disconnect the power from the breadboard and wire the amplifier circuit of Figure 10-10. Depending on how you look at it, it can be seen either as (1) an inverting amplifier with its inverting input grounded via the 10 kΩ resistor, or (2) a noninverting amplifier whose noninverting input is grounded directly.

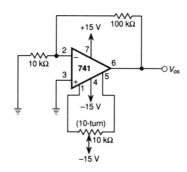

Fig. 10-10. Circuit for nulling output offset voltage.

Step 9

Apply power to the breadboard and measure the output voltage (V_{os}) with a DC voltmeter. Depending on the setting of the 10-kΩ potentiometer, the output offset voltage may be of either polarity and may range from several millivolts on up to the supply voltage.

Step 10

Now vary the setting of the potentiometer. You should observe that the output offset voltage changes. Now carefully adjust the potentiometer until the output reads zero, or at least is within ±2 mV of zero. The op-amp is now effectively balanced so that the output voltage is zero when the input voltage is zero. Once adjusted, the potentiometer is left connected in the circuit at this setting.

Step 11

Turn off and disconnect the power from the breadboard. Wire the circuit shown in the schematic of Figure 10-11a to measure the op-amp's slew rate. Set your oscilloscope for the following approximate settings:

Channel 1: 5 V/division, AC coupling
Channel 2: 1 V/division, AC coupling
Time base: 10 μs/division

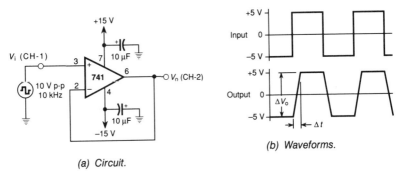

Fig. 10-11. Slew rate measurement.

Step 12

Apply power to the breadboard and adjust the square wave input signal at 10 volts peak-to-peak with a frequency of 10 kHz. The output signal should have a trapezoid shape, as shown in Figure 10-11b. If the op-amp was ideal, having an infinite slew rate, the output signal would look *exactly* the same as the input at very high frequencies. Otherwise, it takes a finite amount of time for the large-signal amplifier to switch from one voltage extreme to the other.

Step 13

Measure the peak-to-peak output voltage (ΔV) and record this value in Table 10-10.

Table 10-10.

Measured peak-to-peak voltage (ΔV)	V
Measured time interval (Δt)	μs
Calculated slew rate	V/μs

Step 14

Measure the time (Δt) that it takes for the output voltage to swing either from its minimum to its maximum value, or vice versa, and record this value in Table 10-10. From these measurements, calculate the slew rate (ΔV/Δt) for your particular 741 op-amp and record your result in the table.

For the 741 op-amp, the slew rate is typically rated at 0.5 V/μs.

Step 15 (Optional)

Disconnect the power and signal generator from the breadboard. Remove the 741 op-amp and replace it with an LM318 device. (The pin connections are exactly the same.) Apply power to the breadboard and repeat Steps 12, 13, and 14. Do you notice any difference between the slew rate of the 741 and LM318?

You should find that the slew rate for the LM318 is much higher than that for a 741. Typically, the slew rate for the LM318 is 70 V/μs.

EXPERIMENT 5—THE DIFFERENTIATOR AND INTEGRATOR

Purpose

The purpose of this experiment is to demonstrate the operation of the differentiator and integrator using op-amps.

Schematic of Circuit

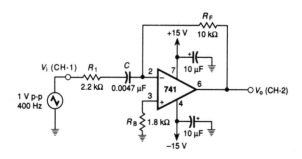

Fig. 10-12. Differentiator for Experiment 5.

Formulas

Differentiator
  Output voltage

$$V_o = - R_F C \frac{\Delta V_i}{\Delta t} \qquad \text{(Eq. 10-15)}$$

Low-frequency response

$$f_1 = \frac{1}{2\pi R_S C} \qquad \text{(Eq. 10-16)}$$

when $f < f_1$, circuit acts as a differentiator
when $f > f_1$, circuit approaches an inverting amplifier with a voltage gain of $-R_F/R_S$.

Integrator
  Output voltage

$$\Delta V_o = - \frac{1}{R_1 C} V_i \Delta t \qquad \text{(Eq. 10-17)}$$

Low-frequency response

$$f_1 = \frac{1}{2\pi R_F C} \qquad \text{(Eq. 10-18)}$$

when $f > f_1$, circuit acts as an integrator
when $f < f_1$, circuit approaches an inverting amplifier with a voltage gain of $-R_F/R_1$.

For minimum output offset due to input bias currents

$$R_B = \frac{R_F R_1}{R_F + R_1} \qquad \text{(Eq. 10-19)}$$

Step 1

Wire the differentiator circuit shown in the schematic of Figure 10-12 and set your oscilloscope to the following approximate settings:

Channel 1: 0.5 V/division, DC coupling
Channel 2: 0.05 V/division, DC coupling
Time base: 0.5 ms/division

Step 2

Apply power to the breadboard and adjust the peak-to-peak voltage of the input triangle wave at 1 V and its frequency to 400 Hz.

You should observe that the output signal is a square wave. Because of the inverting nature of the circuit, the output is positive when the slope of the input triangle wave is negative, while the output is negative when the slope of the input triangle wave is positive.

Step 3

Change the oscilloscope's time base to 0.2 ms/division and Channel 2 to 0.1 V/division. Then adjust the frequency of the input triangle wave to 1 kHz. You should find that the peak-to-peak output voltage of the differentiator increases with increasing input frequency.

Step 4

Now change the input frequency to 30 kHz. Adjust the time base to 10 µs/division and Channel 2 to 2 V/division. What does the output signal look like?

You should now observe that the output signal looks like a triangle wave with phase shift of 180°. Why?

Above approximately 15.4 kHz, the circuit ceases to act as a differentiator since the reactance of the 0.0047-µF capacitor is now less than the 2.2-kΩ resistor (R_S). Above this frequency, the circuit functions like that of an inverting amplifier having a voltage gain of $-R_F/R_S$. Because of the frequency response of the op-amp, there may be some distortion so that the output triangle wave may not look as sharp as the input.

Step 5

Measure the peak-to-peak output voltage and determine the voltage gain. How does it compare to that of an inverting amplifier?

Step 6

Turn off and disconnect the power and input signal from the breadboard and wire the integrator circuit shown in the schematic of Figure 10-13. Set your oscilloscope to the following approximate settings:

    Channels 1 and 2: 0.5 V/division, DC coupling
    Time base: 20 µs/division

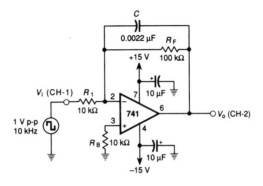

Fig. 10-13. Integrator for Experiment 5.

Step 7

Apply power to the breadboard and adjust the peak-to-peak voltage of the input square wave at 1 V and its frequency to 10 kHz. You should observe that the output signal is a triangle wave.

Because of the inverting nature of the circuit, the slope of the output is positive when the input square wave is negative, while the slope of the output is negative when the input square wave is positive.

Step 8

Change the oscilloscope's time base to 50 μs/division and Channel 2 to 1 V/division. Then adjust the input frequency to 4 kHz. You should find that the integrator's peak-to-peak output voltage increases with decreasing input frequency.

Step 9

Now change the input frequency to 100 Hz. Adjust the time base to 2 ms/division and Channel 2 to 5 V/division. What does the output signal look like?

You should now notice that the output signal looks like a square wave with a phase shift of 180°. Why?

Below approximately 724 Hz, the circuit ceases to act as an integrator since the reactance of the 0.0022-μF capacitor is now greater than the resistance of the 100-kΩ resistor (R_F). Below this frequency the circuit functions like that of an inverting amplifier having a voltage gain of $-R_F/R_1$.

Step 10

Measure the peak-to-peak output voltage and determine the voltage gain. How does it compare to that of an inverting amplifier?

EXPERIMENT 6—SINGLE SUPPLY BIASED INVERTING AC AMPLIFIER

Purpose

The purpose of this experiment is to demonstrate (1) the effect of varying the bias resistors on the quiescent DC output voltage, and (2) the operation of an AC inverting amplifier powered by a single supply voltage.

Schematic of Circuit

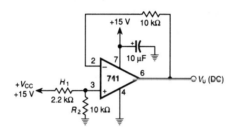

Fig. 10-14. Biasing circuit for Experiment 6.

Formulas

Quiescent DC output voltage

$$V_o(DC) = \left(\frac{R_2}{R_1 + R_2}\right) V_{CC} \qquad \text{(Eq. 10-20)}$$

Usually

$$V_o(DC) = \frac{V_{CC}}{2}$$

when $R_2 = R_1$

Inverting amplifier
  Closed-loop voltage gain

$$A_{CL} = -\frac{R_F}{R_1} \quad (180° \text{ phase shift}) \quad \text{(Eq. 10-21)}$$

Coupling capacitors (low-frequency response)

$$C_1 = \frac{1}{2\pi f_c R_1} \quad \text{(Eq. 10-22)}$$

$$C_2 = \frac{1}{2\pi f_c R_L} \quad \text{(Eq. 10-23)}$$

Step 1

Wire the circuit shown in the schematic of Figure 10-14 and apply power to the breadboard. Make sure that pin 4 of the op-amp is now connected to ground. With your DC voltmeter, measure the power supply voltage (V_{CC}) and record this value in Table 10-11.

Table 10-11.

Supply Voltage (V_{CC})			V
R_1	V_B	V_o	
2.2 kΩ			
4.7 kΩ			
6.8 kΩ			
10 kΩ			
33 kΩ			
47 kΩ			
68 kΩ			
100 kΩ			

Step 2

Separately measure the DC voltages at pins 3 (V_B) and 6 (V_o) of the op-amp with respect to ground and record these values in Table 10-11.

Step 3

Vary resistor R_1 to the eight remaining values given in Table 10-11. Each time you change R_1, first turn off and disconnect the power from the breadboard. Record each measured value and compare it with the expected value (Eq. 10-20). At what resistance value is the DC output voltage approximately one-half the supply voltage measured in Step 1?

You should find that the quiescent output voltage is one-half the supply voltage when $R_1 = R_2$, or 10 kΩ. In general, for proper operation from a single supply voltage, the quiescent DC output voltage is made equal to one-half the supply voltage.

Step 4

Turn off and disconnect the power from the breadboard and wire the inverting amplifier circuit shown in Figure 10-15. Set your oscilloscope to the following approximate settings:

    Channels 1 and 2: 0.1 V/division, DC coupling
    Time base: 1 ms/division

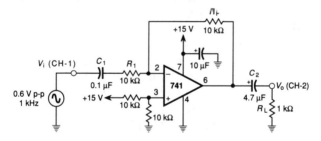

Fig. 10-15. Single supply inverting amplifier for Experiment 6.

Step 5

Apply power to the breadboard and adjust the input signal at 0.6 V peak-to-peak and its frequency to 1 kHz. Is there any difference between the amplifier's input and output signals? You should observe that the output amplitude is the same as the input although it is 180° out of phase with the input.

Step 6

Measure the peak-to-peak output voltage and calculate the voltage gain. Record these values in Table 10-12 and compare them with the expected values (Eq. 10-21). You should calculate a closed-loop voltage gain of 1, since both signals are the same voltage.

Table 10-12.

R_F	Measured V_o	Measured Gain	Expected Gain
10 kΩ			
22 kΩ			
47 kΩ			
68 kΩ			
100 kΩ			
4.7 kΩ			

Step 7

Now transfer the Channel 2 probe to the output pin of the op-amp (pin 6). Are there any differences between the output signal at pin 6 and the output signal across the 1-kΩ load?

You should have observed that (1) the amplitude of both signals is the same, and (2) that both signals are each 180° out of phase with the input signal. Most importantly, however, the output signal at pin 6 of the op-amp is approximately one-half the supply voltage higher than the signal across the load. This is because the output coupling capacitor (C_2) AC couples the signal at the output of the op-amp to the load removing the DC quiescent voltage.

Step 8

Keeping the input signal constant at 0.6 V, change resistor R_F to the remaining five values given in Table 10-12. For each resistance, determine the amplifier's closed-loop voltage gain and compare each value with the expected value (Eq. 10-21). Record all results in the table.

EXPERIMENT 7—THE NORTON AMPLIFIER

Purpose

The purpose of this experiment is to demonstrate (1) how to bias an LM3900 current differencing (Norton) amplifier, and (2) its operation as noninverting and inverting AC amplifiers.

Schematic of Circuit

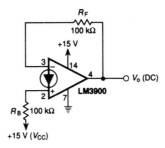

Fig. 10-16. Norton amplifier bias circuit for Experiment 7.

Formulas

Quiescent DC output voltage

$$V_o(DC) = \left(\frac{R_F}{R_B}\right) V_{CC} \quad \text{(Eq. 10-24)}$$

Usually

$$V_o(DC) = \frac{V_{CC}}{2}$$

when $R_B = 2R_F$

Noninverting amplifier
Closed-loop voltage gain

$$A_{CL} = \frac{R_F}{R_1} \quad \text{(Eq. 10-25)}$$

Coupling capacitors (low-frequency response)

$$C_1 = \frac{1}{2\pi f_c R_1} \quad \text{(Eq. 10-26)}$$

$$C_2 = \frac{1}{2\pi f_c R_L} \quad \text{(Eq. 10-27)}$$

Inverting amplifier
Closed-loop voltage gain

$$A_{CL} = -\frac{R_F}{R_1} \quad (180° \text{ phase shift}) \quad \text{(Eq. 10-28)}$$

Coupling capacitors (low-frequency response)

$$C_1 = \frac{1}{2\pi f_c R_1} \quad \text{(Eq. 10-29)}$$

$$C_2 = \frac{1}{2\pi f_c R_L} \quad \text{(Eq. 10-30)}$$

Step 1

Wire the bias circuit shown in the schematic of Figure 10-16 and apply power to the breadboard. With your DC voltmeter, measure the power supply voltage (V_{CC}) and record this value in Table 10-13.

Table 10-13.

Supply Voltage (V_{CC})		V
R_B	Measured $V_o(DC)$	Expected $V_o(DC)$
100 kΩ		
150 kΩ		
200 kΩ		
270 kΩ		
330 kΩ		
390 kΩ		
470 kΩ		

Step 2

Next, measure the DC output voltage at pin 4 of the LM3900 while varying resistor R_B to the remaining six values given in Table 10-13. Each time you change R_B, first disconnect the power and signal generator from the breadboard. Then record each measured value in the table and compare it with the expected value (Eq. 10-24). At what resistance value for R_B is the quiescent DC output voltage approximately one-half the supply voltage measured in Step 1?

You should find that the quiescent output voltage is one-half the supply voltage when $R_B = 2R_F$, or 200 kΩ. In fact, the output voltage is about 0.7 to 0.8 V less than the supply voltage. Although this is slightly less than the supply voltage, it is nevertheless the maximum output voltage that can be obtained from the op-amp. It is called the *saturation voltage* (V_{SAT}). In reality, V_{SAT} should be used in place of V_{CC} in all calculations.

Step 3

Turn off and disconnect the power from the breadboard and wire the noninverting amplifier circuit shown in Figure 10-17. Set your oscilloscope to the following approximate settings:

Channels 1 and 2: 0.2 V/division, DC coupling
Time base: 1 ms/division

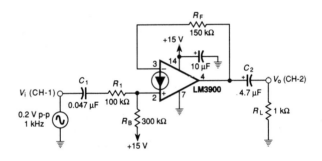

Fig. 10-17. Noninverting Norton amplifier for Experiment 7.

Step 4

Apply power to the breadboard and adjust the input signal at 0.2 V peak-to-peak and its frequency to 1 kHz. Is there any difference between the amplifier's input and output signals?

The only difference between these two signals is that the output signal is *larger* than the input, while both signals are in phase.

Step 5

Measure the peak-to-peak output voltage (V_o) and calculate the voltage gain. Record these values in Table 10-14 and compare them with the expected values (Eq. 10-25).

Table 10-14.

Parameter	Noninverting Amplifier	Inverting Amplifier
Applied input voltage (V_i)		
Measured output voltage (V_o)		
Measured voltage gain		
Expected voltage gain		

Step 6

Now transfer the Channel 2 probe to the output pin of the Norton amplifier (pin 4). Are there any differences between the output signal at pin 4 and the output signal across the 1-kΩ load?

You should have observed that (1) the amplitude of both signals is the same, and (2) that both signals are in phase with the input signal. Most importantly, however, the output signal at pin 4 of the op-amp is approximately one-half the supply voltage higher than the signal across the load. This is because the output coupling capacitor (C_2) AC couples the signal at the output of the op-amp to the load without the DC quiescent voltage.

Step 7

Turn off and disconnect the power and signal generator from the breadboard and wire the inverting amplifier circuit shown in Figure 10-18. Set your oscilloscope to the following approximate settings:

    Channels 1 and 2: 0.1 V/division, DC coupling
    Time base: 1 ms/division

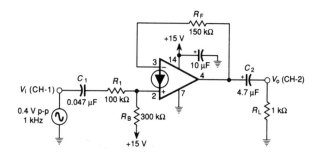

Fig. 10-18. Inverting Norton amplifier for Experiment 7.

Step 8

Apply power to the breadboard and adjust the input signal at 0.4 V peak-to-peak and its frequency to 1 kHz. Are there any differences between the amplifier's input and output signals?

You should have observed that the output signal is larger than the input and inverted with respect to the input (180° out of phase).

Step 9

Measure the peak-to-peak output voltage and calculate the voltage gain. Record these values in Table 10-14 (Step 5) and compare them with the expected values (Eq. 10-28).

Step 10

Now transfer the Channel 2 probe to the output pin of the Norton amplifier (pin 4). Are there any differences between the output signal at pin 4 and the output signal across the 1-kΩ load?

You should have observed that (1) the amplitude of both signals is the same, and (2) that both signals are 180° out of phase with the input signal. Most importantly however, the output signal at pin 4 of the op-amp is approximately one-half the supply voltage *higher* than the signal across the load. This is because the output coupling capacitor (C_2) AC couples the signal at the output of the op-amp to the load *without* the DC quiescent voltage.

EXPERIMENT 8—PRECISION HALF- AND FULL-WAVE RECTIFIERS

Purpose

The purpose of this experiment is to demonstrate the operation of precision half- and full-wave rectifiers.

Schematic of Circuit

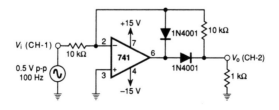

Fig. 10-19. Precision half-wave rectifier for Experiment 8.

Formulas

Half-wave rectifier DC (average) output voltage

$$V_{DC} = 0.318\ V_p \qquad \text{(Eq. 10-31)}$$

Precision full-wave rectifier DC output voltage

$$V_{DC} = 0.636\ V_p \qquad \text{(Eq. 10-32)}$$

Step 1

Wire the precision half-wave rectifier circuit shown in Figure 10-19, paying careful attention to the polarity of the 1N4001 diodes.

Step 2

Next, set your oscilloscope to the following approximate settings:

Channels 1 and 2: 1 V/division, DC coupling
Time base: 2 ms/division, chopped triggering

Step 3

Turn on the function generator and adjust the sine-wave output signal at 6 V peak-to-peak and its frequency to 100 Hz. You should see the rectifying action when comparing the input signal with the signal across the 1-kΩ load. When the input signal goes positive, the output voltage also goes positive. When the input goes negative, the output is zero. From the oscilloscope, measure the peak voltages of the half-wave rectified output and record this value in Table 10-15.

Step 4

With a DC voltmeter, measure the DC voltage across the load resistor, recording your result in Table 10-15. Compare this with the expected value using the peak rectified voltage (V_P) across the 1-kΩ load (Eq. 10-31).

Step 5

Using the remaining seven peak-to-peak input voltage levels listed in Table 10-15, again measure the peak rectified and DC voltage levels across the load, recording your results.

Table 10-15.

Applied Peak-to-peak AC Voltage	Measured Peak Load Voltage	Measured DC Load Voltage	Expected DC Load Voltage
0.5 V			
0.8 V			
1.0 V			
2.0 V			
4.0 V			
8.0 V			
14.0 V			

Step 6

Now turn off the power supplies and signal generator. Wire the precision full-wave rectifier circuit shown in Figure 10-20. Turn on the function generator and adjust its sine-wave output signal at 0.5 V peak-to-peak at a frequency of 100 Hz.

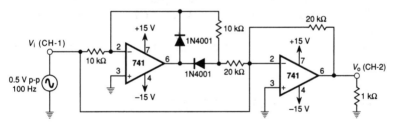

Fig. 10-20. Precision full-wave rectifier for Experiment 8.

When input signal goes positive, the output also goes positive without the loss of voltage due to the voltage drop of the diode's barrier potential. When the input signal goes negative, the output again goes positive. Measure the peak voltage levels of both the input sine wave and the full-wave rectified output, recording your results in Table 10-16.

Step 7

With a DC voltmeter, measure the DC voltage across the load resistor and record this result in Table 10-16. Compare this with the

Table 10-16.

Applied Peak-to-peak AC Voltage	Measured Peak Load Voltage	Measured DC Load Voltage	Expected DC Load Voltage
0.5 V			
0.8 V			
1.0 V			
2.0 V			
4.0 V			
8.0 V			
14.0 V			

expected value using the peak rectified voltage across the 1-kΩ load (Eq. 10-32).

Step 8

Using the remaining peak-to-peak input voltage levels listed in Table 10-16, again measure the peak rectified and DC voltage levels across the load.

How do the two rectifiers compare at low voltages? At higher voltages?

EXPERIMENT 9—THE PEAK DETECTOR

Purpose

The purpose of this experiment is to demonstrate the operation of a peak detector circuit using a 741 op-amp followed by a JFET-input op-amp acting as a buffer.

Step 1

Wire the circuit shown in the schematic of Figure 10-21 and set your oscilloscope for the following approximate settings:

   Channels 1 and 2: 0.5 V/division, DC coupling
   Time Base: 0.5 ms/division

Schematic of Circuit

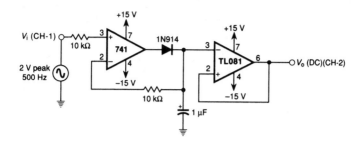

Fig. 10-21. Peak Detector for Experiment 9.

Step 2

Apply power to the breadboard and adjust the output of the signal generator at a peak voltage of 2 V and its frequency to 500 Hz.

Step 3

With the oscilloscope, measure the level of the constant output voltage of the JFET op-amp and record its value in Table 10-17.

You should observe that the DC output voltage is the same as the peak voltage of the input sine wave. The first op-amp circuit is a half-wave rectifier which charges the 1-μF capacitor to the peak value of the input waveform; the JFET op-amp acts as a high input impedance voltage follower.

Table 10-17.

Step	Peak Input Voltage	Measured DC Output Voltage	Remarks
3	2 V		Initial input level
4	2 V		Capacitor shorted
4	2 V		Short removed
5	5 V		Increased input level
6	3 V		Decreased input level
7	3 V		Capacitor shorted
7	3 V		Short removed

Step 4

Momentarily, place a short piece of wire across the 1-µF capacitor and notice what happens to the output voltage. Record the value of the output voltage in Table 10-17.

You should notice that the output drops to zero when the capacitor is shorted, and then charges up to the input signal's current peak value when the wire is removed. This action serves as a simple *reset* function and can be used to reset the circuit at any time.

Step 5

Increase the peak input voltage to 5 V. Measure the output voltage of the JFET op-amp and record its value in Table 10-17. What happens?

You should have now measured a DC voltage of +5 V, which is equal to the positive peak input voltage.

Step 6

Now decrease the peak input voltage to +3 V. Measure the output voltage of the JFET op-amp and record its value in Table 10-17. What happens? Why?

You should find that the DC output voltage did not change and that it is still +5 V. You must remember that a peak detector is a circuit that remembers the peak positive or negative excursion of an input signal for an infinite period of time until it is reset. Since the circuit was not reset prior to decreasing the input signal level from 5 V to 3 V, the circuit then remembered that the maximum or peak input voltage was still 5 V.

In Step 5 when the input level was increased from 2 V to 5 V, the new peak voltage was greater than the old level, and the capacitor charged up to this new level.

Step 7

Momentarily, place a short piece of wire across the 1-µF capacitor and notice what happens to the output voltage. Record the value of the output voltage in Table 10-17.

As in Step 4, you should notice that the output drops to zero when the capacitor shorted, and then charges up to only 2 V, which is the new peak value. For the peak detector to give a true measure of the peak value of the instantaneous input signal, the circuit must first be reset before any reliable measurements can be made.

EXPERIMENT 10—OPERATIONAL AMPLIFIER COMPARATORS

Purpose

The purpose of this experiment is to demonstrate the operation of noninverting and inverting comparator circuits.

Schematic of Circuit

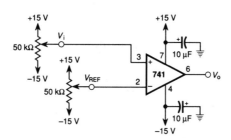

Fig. 10-22. Noninverting comparator for Experiment 10.

Formulas

Noninverting comparator output

$$V_o = V_{+SAT}$$

when $V_i > V_{REF}$

$$V_o = V_{-SAT}$$

when $V_i < V_{REF}$

Inverting comparator output

$$V_o = V_{+SAT}$$

when $V_i < V_{REF}$

$$V_o = V_{-SAT}$$

when $V_i > V_{REF}$

Step 1

Wire the noninverting comparator circuit shown in the schematic of Figure 10-22 and apply power to the breadboard. With your DC voltmeter, adjust the INPUT potentiometer so that the comparator's DC input voltage (V_i) at pin 3 is -10 V with respect to ground.

Step 2

Now adjust the REFERENCE potentiometer so that the comparator's DC reference voltage (V_{REF}) at pin 2 is $+2$ V with respect to ground. Now measure the DC output voltage of the comparator at pin 6. Record this value in Table 10-18.

Table 10-18.

V_{REF}	V_i	V_o
$+2$ V	-10 V	
$+2$ V	$+10$ V	

You should observe that the comparator's output voltage is approximately -14 V, indicating that the comparator's input voltage is less than its reference voltage. As a result, the output voltage of the comparator is equal to the op-amp's negative saturation voltage. Using a negative supply voltage of -15 V, this is approximately -14 V and can vary somewhat from one device to another.

Step 3

Adjust the INPUT potentiometer so that the comparator's input voltage (V_i) at pin 3 is $+10$ V with respect to ground. Again measure the DC output voltage of the comparator at pin 6 and record its value in Table 10-18.

You should observe that the output voltage is approximately $+14$ V, indicating that the comparator's input voltage is greater than its reference voltage. As a result, the output voltage of the comparator is equal to the op-amp's positive saturation voltage. Using a positive supply voltage of $+15$ V, this is approximately $+14$ V.

Step 4

Vary the INPUT potentiometer's setting from one extreme to another while measuring the comparator's DC output voltage. What happens?

You should observe that the output voltage switches between the op-amp's two saturation voltages. The output voltage then depends on the relation of the comparator's input voltage at pin 3 to the reference voltage (pin 2).

Step 5

Turn off and disconnect power from the breadboard and connect pin 3 of the comparator as shown in Figure 10-23a. Using a DC voltmeter, adjust the REFERENCE potentiometer so that the comparator's reference voltage (V_{REF}) at pin 2 is now +1 V with respect to ground. Also, set your oscilloscope for the following approximate settings:

    Channel 1: 1 V/division, DC coupling
    Channel 2: 10 V/division, DC coupling
    Time base: 1 ms/division

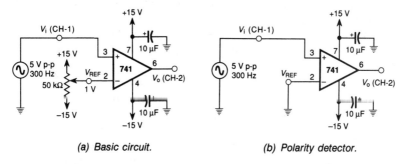

(a) Basic circuit.   (b) Polarity detector.

Fig. 10-23. Noninverting comparator with sine-wave input signal.

Step 6

Apply power to the breadboard and adjust the input voltage at 5 V peak-to-peak and its frequency to 300 Hz. What is the output voltage when the input signal goes positive? When the input goes negative?

You should observe that when the input signal (Channel 1) is greater than the 1-V reference voltage, the output voltage (Channel 2) equals the op-amp's positive saturation voltage. When it is less than the 1-V reference, the output equals the negative saturation voltage. Vary the peak input voltage higher and lower. What happens?

You should observe that the input sine-wave signal is converted into a rectangular pulse train, having the same frequency of the input sine wave. As long as the positive peak voltage of the input signal exceeds the reference voltage, the output voltage equals the op-amp's positive saturation voltage. When it is less than the refer-

203

ence, the output equals the negative saturation voltage. If the peak positive voltage never exceeds the reference voltage, then the output remains at the negative saturation voltage.

Step 7

Set the input sine wave at a peak voltage of 6 V. Vary the REFERENCE potentiometer higher and lower. What happens?

You should observe that the pulse train, while still having the same frequency as the input sine wave, has a duty cycle that varies with the level of the reference voltage.

Step 8

Turn off and disconnect the power and the signal generator voltage from the breadboard. Rewire the op-amp's inverting input (pin 2) directly to ground as shown in Figure 10-23b. Apply power to the breadboard and adjust the input voltage at 5 V peak-to-peak and its frequency to 300 Hz. What happens? Why?

You should observe that the comparator's output signal is now a square wave, which is a pulse train having a 50 percent duty cycle. The reference voltage is now zero, so that the input signal's peaks are the same in either direction from ground reference. This type of circuit is often referred to as either a *sine-to-square wave converter*, *polarity detector*, or *zero crossing detector*. In all cases, the comparators are noninverting comparators as the polarity of the comparator's input and output signals are the same when the input signal is greater than the reference voltage.

Step 9

Disconnect the power and input signal from the breadboard. Then wire the inverting comparator circuit shown in the schematic of Figure 10-24 and apply power to the breadboard. With a DC voltmeter, adjust the INPUT potentiometer so that the comparator's input voltage (V_i) at pin 2 is −10 V with respect to ground.

Step 10

Adjust the REFERENCE potentiometer so that the comparator's reference voltage (V_{REF}) at pin 3 is +2 V with respect to ground. Now measure the DC output voltage of the comparator at pin 6. Record this value in Table 10-19.

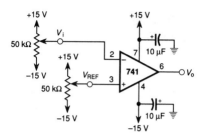

Fig. 10-24. Inverting comparator for Experiment 10.

Table 10-19.

V_{REF}	V_i	V_o
+2 V	−10 V	
+2 V	+10 V	

You should observe that the comparator's output voltage is approximately +14 V, indicating that the comparator's input voltage is less than its reference voltage. As a result, the output voltage of the inverting comparator is equal to the op-amp's positive saturation voltage.

Step 11

Adjust the INPUT potentiometer so that the comparator's input voltage (V_i) at pin 2 is +10 V with respect to ground. Now measure the DC output voltage of the comparator at pin 6. Again record this value in Table 10-19.

You should observe that the comparator's output voltage is now approximately −14 V, indicating that the inverting comparator's input voltage is greater than its reference voltage. As a result, the output voltage of the comparator is equal to the op-amp's negative saturation voltage.

Step 12

Vary the INPUT potentiometer's setting from one extreme to another while measuring the comparator's DC output voltage. What happens?

You should observe that the output voltage switches between the op-amp's two saturation voltages. The output voltage then depends on the relation of the comparator's input voltage at pin 2 to the reference voltage (pin 3).

Step 13

Turn off and disconnect power from the breadboard and connect pin 2 of the comparator to the output of the signal generator as shown in Figure 10-25a. Using a DC voltmeter, adjust the REFERENCE potentiometer so that the comparator's reference voltage (V_{REF}) at pin 3 is now +1 V with respect to ground. Also, set your oscilloscope for the following approximate settings:

Channel 1: 1 V/division, DC coupling
Channel 2: 10 V/division, DC coupling
Time base: 1 ms/division

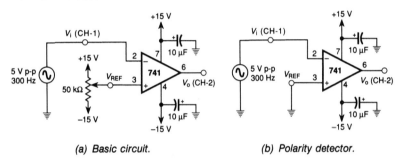

(a) Basic circuit.   (b) Polarity detector.

Fig. 10-25. *Inverting comparator with sine-wave input signal.*

Step 14

Apply power to the breadboard and adjust the input voltage at 5 V peak-to-peak and the frequency at 300 Hz. What is the output voltage when the input signal goes positive? When the input goes negative?

You should observe that when the input signal (Channel 1) is greater than the 1-V reference voltage, the output voltage (Channel 2) equals the op-amp's *negative* saturation voltage. When it is less than the 1-V reference, the output equals the *positive* saturation voltage. Vary the peak input voltage higher and then lower. What happens?

You should observe that the input sine-wave signal is converted to a rectangular pulse train, having the same frequency of the input sine wave. As long as the positive peak voltage of the input signal exceeds the reference voltage, the output voltage equals the op-amp's negative saturation voltage. When it is less than the reference, the output equals the positive saturation voltage.

Step 15

Disconnect the power and the signal generator voltage from the breadboard. Rewire the op-amp's noninverting input (pin 3) directly to ground as shown in Figure 10-25b. Apply power to the breadboard and adjust the input voltage at 5 V peak-to-peak and the frequency at 300 Hz. What happens? Why?

You should observe that the comparator's output signal is now a square wave, having a 50 percent duty cycle. The reference voltage is now zero, so that the input signal's peaks are the same in either direction from the reference. This is also a sine-to-square wave converter.

EXPERIMENT 11—THE PHASE-SHIFT OSCILLATOR

Purpose

The purpose of this experiment is to demonstrate the design and operation of a phase-shift oscillator using an op-amp.

Schematic of Circuit

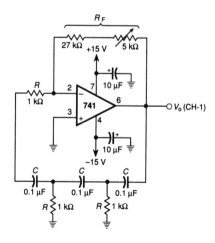

Fig. 10-26. *Phase-shift oscillator for Experiment 11.*

Formulas

Output frequency

$$f_o = \frac{1}{2\pi RC\sqrt{6}}$$ (Eq. 10-33)

Amplifier closed-loop gain for oscillation

$$\frac{R_F}{R} = 29$$ (Eq. 10-34)

Step 1

Wire the phase-shift oscillator circuit shown in the schematic of Figure 10-27, and set your oscilloscope to the following approximate settings:

  Channel 1: 5 V/division, AC coupling
  Time base: 0.5 ms/division

Step 2

After you have checked all connections, apply the power supply connections to the breadboard. Depending on the initial setting of the 5-kΩ potentiometer, the circuit may or may not be oscillating when power is applied. If a sine wave is not displayed on the oscilloscope, carefully adjust the setting of the 5-kΩ potentiometer until a sine wave starts to appear on oscilloscope's display.

If you continue to increase the resistance of the potentiometer you should observe that the waveform becomes distorted, the peaks are clipped, and its output frequency becomes lower. Adjust the potentiometer to the point at which the best looking sine wave appears on the oscilloscope.

On the other hand, if a sine wave is seen when power is applied to the breadboard, carefully adjust the potentiometer to the point at which the best looking sine wave appears on the oscilloscope.

Step 3

Using the oscilloscope's time base, measure the oscillator's output frequency and record its value in Table 10-20. Compare this value with the expected frequency (Eq. 10-33).

Table 10-20.

Parameter	Measured Value	Expected Value
Output frequency		
R_F		29 kΩ

Step 4

Disconnect the power from the breadboard and carefully remove and measure the total series resistance of the 27-kΩ resistor and of the 5-kΩ potentiometer that produced oscillation. Record this value in Table 10-20.

At the oscillation frequency set by the three RC networks, $1/29$ of the output signal is fed back to the inverting input of the op-amp. For the loop gain to be 1, the closed-loop voltage gain of the inverting amplifier section must then be 29, which means that R_F must be 29 times larger than R (1 kΩ).

EXPERIMENT 12—2ND-ORDER LOW- AND HIGH-PASS BUTTERWORTH ACTIVE FILTERS

Purpose

The purpose of this experiment is to compare the design and characteristics of Butterworth Sallen-Key 2nd-order low- and high-pass active filters.

Schematic of Circuit

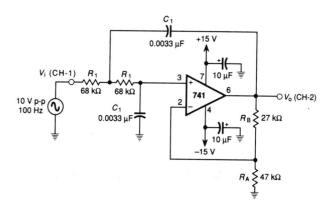

Fig. 10-27. 2nd-order low-pass filter for Experiment 12.

Formulas

Basic requirements

$$R_1 = R_2$$
$$\text{and} \quad \text{(Eq. 10-35)}$$
$$C_1 = C_2$$

$$R_B = 0.586 \, R_A \quad \text{(Eq. 10-36)}$$

Passband voltage gain

$$A_{PB} = 1 + \frac{R_B}{R_A} \quad \text{(Eq. 10-37)}$$

Cutoff frequency

$$f_c = \frac{1}{2\pi R_1 C_1} \quad \text{(Eq. 10-38)}$$

dB frequency response

$$A_{dB} = 20 \, \log(V_o/V_i) \quad \text{(Eq. 10-39)}$$

$$A_{dB} = 20 \, \log \frac{1.586}{\left[1 + (f_{in}/f_c)^4\right]^{1/2}} \quad (low\text{-}pass \; filter) \quad \text{(Eq. 10-40)}$$

$$A_{dB} = 20 \, \log \frac{1.586}{\left[1 + (f_c/f_{in})^4\right]^{1/2}} \quad (high\text{-}pass \; filter) \quad \text{(Eq. 10-41)}$$

Step 1

Wire the low-pass filter circuit shown in the schematic of Figure 10-27 and set your oscilloscope for the following approximate settings:

Channels 1 and 2: 2 V/division
Time base: 1 ms/division, AC coupling

Step 2

Apply power to the breadboard and adjust the input voltage at 10 V peak-to-peak and its frequency to 100 Hz. You should make this voltage setting as accurate as you can.

Step 3

With the resistor and capacitor values used in this circuit, what do you expect the cutoff frequency to be?

From Equation 10-38, the cutoff frequency is approximately 700 Hz.

Step 4

With the input frequency set at 100 Hz, what is the peak-to-peak output voltage, and how does it compare with the expected value?

At frequencies well below the filter's cutoff frequency, the passband voltage gain is controlled by resistors R_A and R_B, and is equal to $1 + R_B/R_A$ (Eq. 10-37). For a 2nd-order Butterworth response, the output voltage should ideally be equal to 1.586 times larger than the input level. This is equivalent to a gain of +4 dB.

You should also observe that both the input and output signals are essentially in phase.

Step 5

Now vary the generator's frequency (f_{in}), keeping the input voltage constant at 10 volt peak-to-peak in order to complete the required data in Table 10-21.

Using the dB frequency response formula (Eq. 10-40), calculate the expected dB response using a cutoff frequency of 700 Hz for each measurement. Then plot both your experimental and expected results on the blank graph of Figure 10-28 provided for this purpose.

From your plotted results, you should find that both curves should coincide. How close they coincide will depend on how close the actual resistor and capacitor values are to the values shown in the schematic. If nothing else, the two curves should essentially be parallel.

Step 7

You should notice that the filter's gain at low frequencies is essentially constant (i.e., the passband) before which it decreases at a linear rate with increasing input frequency. This linear decrease in gain as a function of frequency is termed the *rolloff*. To determine the rolloff of your filter, you must calculate the slope of the line.

Table 10-21.

Frequency	V_o	Measured dB Response	Expected dB Response
100 Hz			
200 Hz			
400 Hz			
600 Hz			
700 Hz			
800 Hz			
1 kHz			
2 kHz			
4 kHz			
8 kHz			
10 kHz			
Rolloff		dB/decade	

From the data in Table 10-21, subtract the dB gain measured at 1 kHz from that measured at 10 kHz. The frequency difference from 1 kHz to 10 kHz is 1 decade (i.e, a factor of 10). The rolloff is the difference in the dB gain over a one decade frequency range. From your measurements, calculate the filter's rolloff and compare it with what you should expect for a 2nd-order Butterworth low-pass filter.

You should find that the 2nd-order low-pass filter's rolloff is about −40 dB/decade.

Step 8

The filter's cutoff frequency is that frequency where the dB frequency response is 3 dB less than the dB passband gain. This is equivalent to an output voltage that is 0.707 times the input voltage of the filter. From your graph, estimate the filter's cutoff frequency and compare it with the expected value calculated in Step 3.

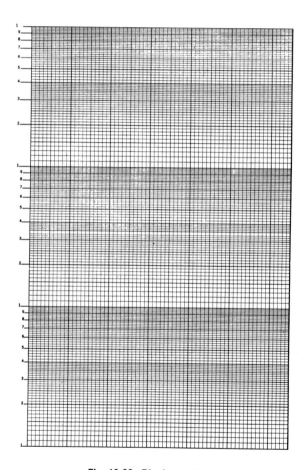

Fig. 10-28. Blank graph.

Step 9

Turn off and disconnect the power to the breadboard and wire the high-pass filter circuit shown in the schematic of Figure 10-29.

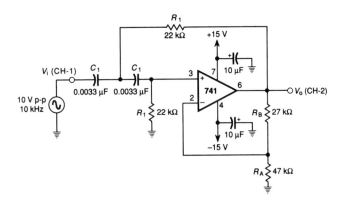

Fig. 10-29. 2nd-order high-pass filter for Experiment 12.

Step 10

Set your oscilloscope for the following approximate settings:

    Channels 1 and 2: 2 V/division
    Time base: 0.1 ms/division, AC coupling

Step 11

Apply power to the breadboard and adjust the input voltage to 10 V peak-to-peak at a frequency of 10 kHz. You should make this voltage setting as accurate as you can.

Step 12

With the resistor and capacitor values used in this circuit, what do you expect the cutoff frequency to be?
    From Equation 4, the cutoff frequency is approximately 2.2 kHz.

Step 13

With the input frequency set at 10 kHz, what is the peak-to-peak output voltage, and how does it compare with the expected value?
    At frequencies well above the filter's cutoff frequency, the passband gain is controlled by resistors R_A and R_B, and is equal to 1 + R_B/R_A. For a 2nd-order Butterworth response, the output voltage should ideally be equal to 1.586 times larger than the input level. This is then equivalent to a gain of 4 dB. You should also observe that both the input and output signals are essentially in phase.

Step 14

Now vary the generator's frequency (f_{in}), keeping the input voltage constant at 10 volt peak-to-peak in order to complete the required data in Table 10-22.

Table 10-22.

Frequency	V_o	Measured dB Response	Expected dB Response
100 Hz			
200 Hz			
400 Hz			
600 Hz			
700 Hz			
800 Hz			
1 kHz			
2 kHz			
4 kHz			
8 kHz			
10 kHz			
Rolloff		dB/decade	

Using the dB frequency response formula (Eq. 10-41), calculate the expected dB response using a cutoff frequency of 2.2 kHz for each measurement. Then plot both your experimental and expected results on the blank graph of Figure 10-30 provided for this purpose.

Step 15

From your plotted results, you should find that both curves should coincide. How close they coincide will depend on how close the actual resistor and capacitor values are to the values shown in the schematic. If nothing else, the two curves should essentially be parallel.

215

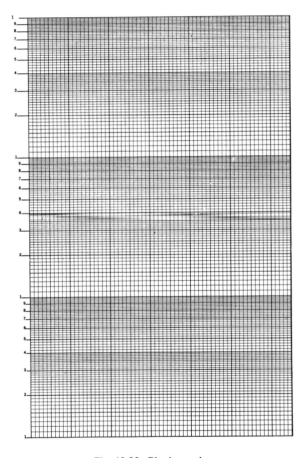

Fig. 10-30. Blank graph.

Step 16

You should notice that the filter's gain at low frequencies increases at a linear rate with increasing input frequency and, beyond some point, levels off and remains essentially constant.

To determine the rolloff of your high-pass filter, subtract the dB gain measured at 100 Hz from that measured at 1 kHz (1 decade). From your measurements, what is the filter's rolloff, and how does it compare with what you should expect for a 2nd-order Butterworth high-pass filter?

You should find that the rolloff for the 2nd-order high-pass filter is about +40 dB/decade.

EXPERIMENT 13—ACTIVE BANDPASS AND NOTCH FILTERS

Purpose

The purpose of this experiment is to compare the design and characteristics of a multiple-feedback active bandpass filter with that of a notch filter.

Schematic of Circuit

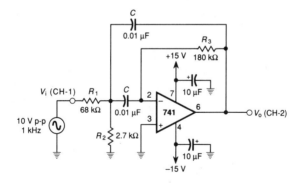

Fig. 10-31. Bandpass filter for Experiment 13.

Formulas

Center (or notch) frequency

$$f_o = \left(\frac{1}{2\pi C}\right)\left(\frac{R_1 + R_2}{R_1 R_2 R_3}\right)^{1/2} \qquad \text{(Eq. 10-42)}$$

217

where

$$R_1 = \frac{Q}{2\pi C f_o A_o}$$

$$R_2 = \frac{Q}{2\pi C f_o (2Q^2 - A_o)}$$

$$R_3 = \frac{Q}{\pi C f_o}$$

Center frequency voltage gain (bandpass filter)

$$A_o = \frac{R_3}{2R_1} \quad (A_o \text{ must be less than } 2Q^2) \quad \text{(Eq. 10-43)}$$

Center (or notch) frequency from upper and lower 3-dB frequencies

$$f_o = (f_L f_H)^{1/2} \quad \text{(Eq. 10-44)}$$

Quality factor

$$Q = \frac{f_o}{f_H - f_L} \quad \text{(Eq. 10-45)}$$

$$Q = \frac{f_o}{BW} \quad \text{(Eq. 10-46)}$$

Step 1

Wire the circuit shown in the schematic of Figure 10-31 and set your oscilloscope for the following approximate settings:

Channels 1 and 2: 2 V/division, AC coupling
Time base: 0.2 ms/division

Step 2

Apply power to the breadboard and adjust the output of the signal generator at 10 V peak-to-peak and its frequency to 1 kHz. Now vary the signal generator's frequency to the point where the output voltage of the filter, as displayed on Channel 2 of the oscilloscope, reaches its *maximum* peak-to-peak amplitude. Measure this peak-to-peak output voltage and then calculate the center frequency dB voltage gain, 20 log (V_o/V_i), and record your data in Table 10-23.

Table 10-23.

Input voltage (V_i) at f_o	V
Output voltage (V_o) at f_o	V
Center frequency voltage gain (A_o)	dB
Center frequency (f_o)	Hz
Lower 3-dB frequency (f_L)	Hz
Upper 3-dB frequency (f_H)	Hz
3-dB bandwidth (BW)	Hz
Quality factor (Q)	

The measured center frequency voltage gain, which is based on resistors R_1 and R_3, should be about 1.32 or +2.4 dB. If your value is more than 10 percent from this value, either you are not at the filter's center frequency, as evidenced by a maximum output voltage, or the resistors you are using are significantly different from their rated values. You should also observe that the input and output waveforms are *exactly 180° out of phase* at this center frequency.

Step 3

Using your oscilloscope, measure filter's center frequency without disturbing the frequency setting of the signal generator and record this value in Table 10-23. How does this compare with the expected value (Eq. 10-42)?

The bandpass filter's center frequency is based on the values of both capacitors and all three resistors, and it should be about 737 Hz.

Step 4

Determine the filter's bandwidth by measuring both the upper and lower 3-dB frequencies where the peak-to-peak output voltage drops to 0.707 times the value at the center frequency. To do this easily, you should set the signal generator to the filter's center frequency as in Step 2. Then, without changing the input frequency, adjust the signal generator's output voltage so that the peak-to-peak output voltage of the *filter* is 14.1 V. Make this setting as accurate as possible.

Then decrease the signal generator's frequency, and stop at the point where the output voltage drops to 10 V peak-to-peak (14.1 V × 0.707 − 10 V), or 5 vertical divisions. Measure the frequency at this point, called the *lower 3-dB frequency* (f_L), and record this value in Table 10-23.

Step 5

Continue to decrease the input frequency. Does the output voltage increase or decrease?

You should observe that the peak-to-peak output voltage of the bandpass filter *decreases* as the input frequency moves away from the filter's center frequency.

Step 6

Now increase the signal generator's frequency beyond the center frequency, and stop at the point where the filter's peak-to-peak output voltage is again 10 V. Measure the frequency at this point, called the *upper 3-dB frequency* (f_H), and record this value in Table 10-23.

Step 7

Calculate the 3-dB bandwidth by subtracting the lower 3-dB frequency from the upper 3-dB frequency of the filter and record this value in Table 10-23. Using this bandwidth and the center frequency experimentally found in Step 3, calculate the filter's Q, or *quality factor* (Eq. 10-45 or 10-46), and record its value in Table 10-23. Within about 10 to 20 percent, you should determine a filter Q of 4.2. If not, repeat Steps 2 through 6, carefully measuring all voltages and frequencies.

Step 8

From the two measured 3-dB frequencies, calculate the filter's center frequency by taking the *geometric average* (Eq. 10-44). How does your result compare with the value measured in Step 3?

Step 9

Set the input voltage to the filter at 10 V peak-to-peak and vary the signal generator's frequency according to Table 10-24. For each of the 10 frequencies, calculate the dB voltage gain and then plot the dB gain response for all measured frequencies on the blank graph of Figure 10-32 provided for this purpose. Also include the center frequency dB voltage gain obtained in Step 2 as a data point.

Table 10-24.

Frequency	V_o	Measured dB response
100 Hz		
300 Hz		
600 Hz		
700 Hz		
800 Hz		
1 kHz		
2 kHz		
4 kHz		
8 kHz		
10 kHz		

From this graph you should be able to estimate the filter's center frequency, bandwidth, and Q, and favorably compare with those values in Table 10-23 of Step 2.

Step 10

Turn off and disconnect the power to the breadboard and wire the notch filter circuit shown in the schematic of Figure 10-33. Here, the notch filter is composed of two sections. The first section is a multiple-feedback bandpass filter identical to that already used in this experiment. It is followed by a 2-input summing amplifier to create a notch filter. Set your oscilloscope for the following approximate settings:

Channels 1 and 2: 10 V/division, AC coupling
Time base: 0.2 ms/division

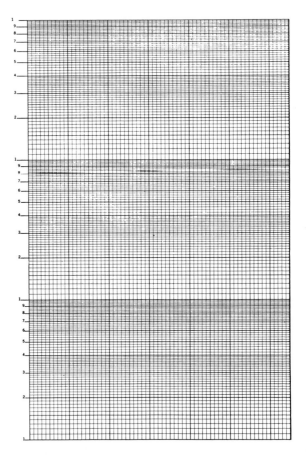

Fig. 10-32. Blank graph.

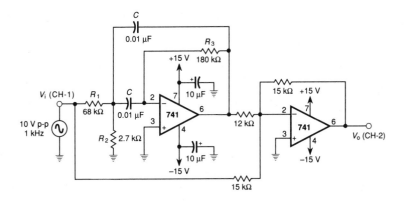

Fig. 10-33. Notch filter for Experiment 13.

Step 11

Apply power to the breadboard and adjust the output of the signal generator at 10 V peak-to-peak and its frequency to 1 kHz.

Step 12

Now slowly decrease signal generator's frequency to the point where the output voltage of the filter, as displayed on Channel 2 of the oscilloscope, reaches its *minimum* peak-to-peak amplitude. You will need to significantly increase the sensitivity of Channel 2 of the oscilloscope to do this. Measure the peak-to-peak output voltage and then calculate the notch frequency voltage gain in decibels, which is sometimes called the *notch depth*, or *depth of null*. Record this value in Table 10-25. You should measure a notch depth of at least −25 dB.

If you do not have a notch depth of at least −25 dB, you probably are not at the filter's notch frequency, or your components are mismatched. You should also observe that the input and output waveforms are *exactly in phase* at this center frequency.

Step 13

Using your oscilloscope, measure the filter's notch frequency without disturbing the frequency setting of the signal generator, recording the value in Table 10-25. How does this compare with the expected value (Eq. 10-42)?

The filter's notch frequency is based on the values of both capacitors and all three resistors, and it should be the same as that for the bandpass filter's center frequency determined in Step 3 (approximately 737 Hz).

Table 10-25.

Input voltage (V_i) at f_o	V
Output voltage (V_o) at f_o	V
Notch depth	dB
Notch frequency (f_o)	Hz
Passband voltage gain	
Lower 3-dB frequency (f_L)	Hz
Upper 3-dB frequency (f_H)	Hz
3-dB bandwidth (BW)	Hz
Quality factor (Q)	

Step 14

Continue to decrease the output frequency. Does the output voltage increase or decrease?

Notice that the peak-to-peak output voltage of the notch filter increases as the input frequency moves away from the filter's notch frequency. Eventually, the output voltage remains constant.

Step 15

Set the frequency of the signal generator at 100 Hz and the oscilloscope's Channel 2 sensitivity to 2 V/division. Measure the peak-to-peak output voltage of the filter and calculate the passband voltage gain, recording this value in Table 10-25. Unlike the bandpass filter, you should find that the passband voltage gain is unity and that the input and output signals are 180° out of phase.

Step 16

Now determine the filter's bandwidth by measuring both the upper and lower 3-dB frequencies where the peak-to-peak output voltage drops to 0.707 times the value in the filter's passband. To do this easily, first set the signal generator to 10 kHz. Then, without changing the input frequency, adjust the signal generator's output voltage so that the output voltage of the filter is 14.1 V. Make this setting as accurate as possible. Then decrease the signal generator's frequency and stop at the point where the output voltage drops to 10 V peak-to-peak (5 vertical divisions). Measure the frequency at this point, called the *upper 3-dB frequency* (f_H) and record this value in Table 10-25.

Step 17

Continue to decrease the input frequency. Does the output voltage increase or decrease?

You should observe that the peak-to-peak output voltage of the bandpass filter *decreases* as the input frequency moves away from the filter's passband.

Step 18

Now continue to decrease the signal generator's frequency beyond the notch frequency and stop at the point where the filter's peak-to-peak output voltage is again 10 V. Measure the frequency at this point, called the *lower 3-dB frequency* (f_L) and record this value in Table 10-25.

Step 19

Calculate the 3-dB bandwidth of the notch filter by subtracting the lower 3-dB frequency from the upper 3-dB frequency, and record this value in Table 10-25. Using this bandwidth and the notch frequency experimentally found in Step 13, calculate the filter's Q, and record its value in the table.

Within about 10 to 20 percent, you should determine a filter Q of 4.2. If not, repeat Steps 11 through 18, carefully measuring all voltages and frequencies.

Step 20

From the two measured 3-dB frequencies, calculate the filter's notch frequency by taking the geometric average (Eq. 10-44).

Step 21

Set the input voltage to the filter at 10 V peak-to-peak and vary the signal generator's frequency according to Table 10-26. For each of the 10 frequencies, calculate the voltage gain in decibels and then plot the dB gain response for all measured frequencies on the blank graph of Figure 10-34 provided for this purpose. Also include the notch frequency dB null depth obtained in Step 13 as a data point.

225

From this graph you should be able to estimate the filter's notch frequency, bandwidth, and Q, and favorably compare with those values in Table 10-25 of Step 12.

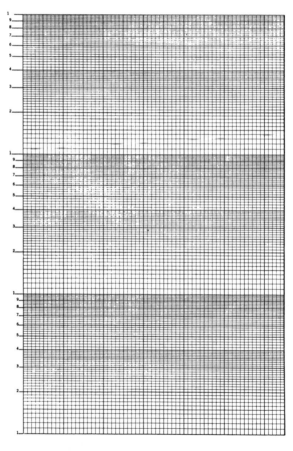

Fig. 10-34. Blank graph.

Table 10-26.

Frequency	V_o	Measured dB Response
100 Hz		
300 Hz		
600 Hz		
700 Hz		
800 Hz		
1 kHz		
2 kHz		
4 kHz		
8 kHz		
10 kHz		

EXPERIMENT 14—THE STATE-VARIABLE FILTER

Purpose

The purpose of this experiment is to demonstrate the operation and characteristics of a state-variable filter using three 741 op-amps.

Schematic of Circuit

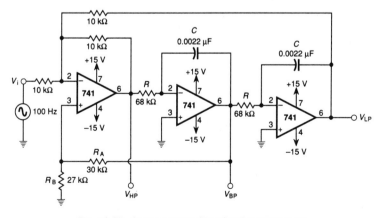

Fig. 10-35. State-variable filter for Experiment 14.

Formulas

Cutoff/center frequency

$$f_c = \frac{1}{2\pi RC} \qquad \text{(Eq. 10-47)}$$

Filter Q

$$R_A = (3Q - 1)R_B \qquad \text{(Eq. 10-48)}$$

$$Q = \frac{f_o}{BW} \qquad \text{(Eq. 10-49)}$$

Passband gain and phase shift

$$A(dB) = 20 \log\left(\frac{V_o}{V_i}\right) \qquad \text{(Eq. 10-50)}$$

Low-pass = -1 (180° out of phase)
High-pass = -1 (180° out of phase)
Bandpass = Q (in phase)

Step 1

Wire the circuit shown in the schematic of Figure 10-35 and set your oscilloscope for the following approximate settings:

Channels 1 and 2: 0.2 V/division, AC coupling
Time base: 1 ms/division

Step 2

Apply power to the breadboard and adjust the output of the signal generator at 1.4 V peak-to-peak (7 vertical divisions) and its frequency to 100 Hz.

Step 3

Now measure the peak-to-peak low-pass output signal (V_{LP}) at pin 6 of the first op-amp. How does this compare with the input signal?
 You should find that the peak-to-peak amplitude of the low-pass output is essentially the same as the input (1.4 V), giving a unity voltage gain. In addition, the low-pass output is inverted with re-

spect to the input so that it is 180° out of phase with the input in the filter's passband.

Step 4

Now set the oscilloscope's time base at 10 μs/division and adjust the input frequency to 10 kHz with a peak-to-peak amplitude at 1.4 V. Now measure the peak-to-peak high-pass output signal (V_{HP}) at pin 6 of the last op-amp. How does this compare with the filter's input signal?

You should find that the peak-to-peak amplitude of the high-pass output is also essentially the same as the input (1.4 V), giving a unity voltage gain. Like the low-pass output, the high-pass output is also inverted with respect to the input so that it is 180° out of phase with the input signal in the filter's passband.

Step 5

Decrease the input frequency until the voltage of the highpass output drops to 1.0 V peak-to-peak (5 vertical divisions), which is approximately 0.707 times the input level or a drop of 3 dB. Measure the input frequency at which this occurs, which is the cutoff frequency of the high-pass filter response, $f_c(HP)$, and record this value in Table 10-27.

Table 10-27.

Parameter	Measured	Calculated
High-pass cutoff frequency [$f_c(HP)$]		
Low-pass cutoff frequency [$f_c(LP)$]		
Center frequency [$f_o(BP)$]		
Quality factor (Q)		
Center frequency gain		

Step 6

Without changing the setting of the signal generator, transfer the oscilloscope's Channel 2 probe from the filter's high-pass output back to the low-pass output and measure the peak-to-peak output voltage. If it is not already 1.0 V, then adjust the input frequency until the low-pass output is also 1.0 V. Measure this frequency, which is the cutoff frequency of the low-pass filter response, $f_c(LP)$, and record this value in Table 10-27. Within 10 percent, both cutoff frequencies should be the same and within the expected value of 1064 Hz (Eq. 10-47).

Step 7

Now connect the oscilloscope's Channel 2 probe to the filter's bandpass output (V_{BP}). Very carefully vary the input frequency higher or lower, stopping at the point at which the output voltage reaches its maximum value. Measure the frequency at which this occurs, called the *center frequency*, $f_o(BP)$, and record its value in Table 10-27.

Normally, the center frequency should be the same as the two cutoff frequencies. However, if the low- and high-pass cutoff frequencies measured in Steps 5 and 6 are different, then the two integrator sections of the state-variable filter are mismatched to some degree.

If this is the case, the center frequency of the bandpass response must then be calculated as the geometric average based on the cutoff frequencies of the low- and high-pass responses

$$f_o(BP) = \left[f_c(LP) \times f_c(HP) \right]^{1/2}$$

Within about 10 percent, the center frequency should be within the expected value of 1064 Hz (Eq. 10-47).

Step 8

Now measure the peak-to-peak output voltage of the bandpass output (V_{BP}). You should observe that it is approximately 1.0 V, indicating a voltage gain of approximately 1.0 V/1.4 V, or 0.71 at the center frequency. This is also numerically equal to the filter's Q. Calculate the Q of the bandpass response (Eq. 10-49) and compare it with the expected value (Eq. 10-48) based on R_A and R_B. Record both values in Table 10-27.

You should realize that the state-variable filter is designed to be used either as low- or high-pass filter with a Butterworth (maximally flat) response, or solely as a high-Q bandpass filter. Although all three responses are available simultaneously, the results are not optimal. From Equation 2, the filter's Q is controlled by resistors

R_A and R_B. For a Butterworth response, Q must equal 0.707, which is not the best for a bandpass filter.

Step 9

Turn off and disconnect the power and signal generator from the breadboard and replace the 30-kΩ resistor (R_A) with a 270-kΩ resistor. Now the Q of the bandpass filter response has been increased to about 3.7 (Eq. 10-48).

Step 10

Repeat Steps 2 through 8, recording your values in Table 10-28. How do your results compare with the expected results?

Table 10-28.

Parameter	Measured	Calculated
High-pass cutoff frequency [f_c(HP)]		
Low-pass cutoff frequency [f_c(LP)]		
Center frequency [f_o(BP)]		
Quality factor (Q)		
Center frequency gain		

Other than a change in the center-frequency voltage gain of the bandpass response as a result of increasing its Q by changing R_A, you should find no other changes.

Step 11

Again disconnect the power and signal generator from the breadboard and replace the 270-kΩ resistor (R_A) with the original 30-kΩ resistor.

Step 12

Change the sensitivity of both Channel 1 and 2 to 1 V/division. Then adjust the level of the filter's input signal at 6 V peak-to-peak and its frequency to 100 Hz.

Step 13

Measure the peak-to-peak voltages of the low-pass (V_{LP}), high-pass (V_{HP}), and bandpass (V_{BP}) outputs. Then calculate the voltage gain in decibels for each output (Eq. 10-50). Record these values in Table 10-29.

Table 10-29.

Frequency	Output Voltage			dB Response		
	V_{LP}	V_{HP}	V_{BP}	A_{LP}	A_{HP}	A_{BP}
100 Hz						
400 Hz						
800 Hz						
900 Hz						
1 kHz						
2 kHz						
4 kHz						
10 kHz						

Step 14

Vary the input frequency to the remaining seven values given in Table 10-29. For each frequency, measure the peak-to-peak voltages of the low-pass (V_{LP}), high-pass (V_{HP}), and bandpass (V_{BP}) outputs. Calculate the voltage gain in decibels for each output and record these values in the table. Make sure that the input level stays constant at 6 V peak-to-peak at all frequencies.

Step 15

Using the results of Table 10-29, plot the frequency responses for all three outputs on the blank graph of Figure 10-36 provided for this purpose. In addition, use the center frequency gain obtained in Step 8 also as a data point for the graph. To make this composite graph easier to view, you might want to use a different color for each curve.

If you have done everything correctly, the low-pass and high-pass response curves should cross at the same frequency where the bandpass response reaches its maximum value. At this point, the two cutoff frequencies and the center frequency have the same value.

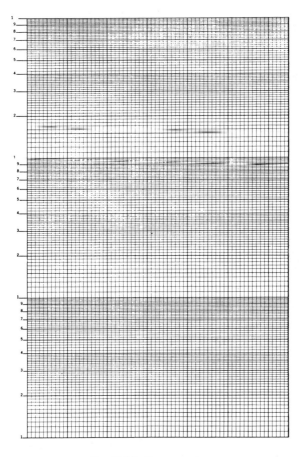

Fig. 10-36. Blank graph.

Glossary

Active filter—Any filter network built around an active device such as a transistor or op-amp to produce gain in the filter's passband.
Bandpass filter—A filter network that allows signals between two frequencies to pass through unaffected, but rejects signal frequencies above and below this range.
Bandwidth—The frequency range between frequencies where the response is 3 dB less than the passband gain.
Bessel filter—A filter response whose low- or high-pass phase shift is linear with frequency.
Butterworth response—A filter response characterized by a flat passband. Also called a *maximally-flat response*.
Center frequency—The median frequency of bandpass and notch filters.
Chebyshev response—A filter response characterized by equal ripples of constant amplitude in the passband. Also called an *equal-ripple response*.
Clipping—The process of preventing an output signal from exceeding a certain positive or negative level, or both. Also called limiting.
Closed-loop voltage gain—The voltage gain of an amplifier circuit with a closed feedback path between the output and the amplifier's input.
Common-mode gain—In a difference amplifier, the ratio of the common-mode output voltage to the common-mode input voltage.
Common-mode rejection ratio (CMRR)—The measure of how well a difference amplifier rejects amplification of a common-mode input voltage, which is based on both the common-mode and differential gains.
Comparator—A circuit that compares an input voltage to a predetermined DC reference voltage.
Cutoff frequency—The frequency of low- and high-pass filters where the response is 3 dB less than the passband gain. Also called the *break, corner, critical,* or *3-dB frequency*.
Difference amplifier—An amplifier circuit whose output voltage is proportional to the voltage difference of its two inputs. Also called a *differential amplifier*.
Differential input voltage—The maximum voltage that can be applied between the inverting and noninverting inputs.
Gain-bandwidth product (GBP)—The product of the closed-loop voltage gain of an op-amp and its corresponding bandwidth.
High-pass filter—A filter network that allows signals above a certain frequency to pass through unaffected, but rejects signal frequencies below this frequency.

Input bias current—The average of the two DC currents that flow into the inverting and noninverting inputs of an op-amp.
Input offset current—The difference of the two input bias currents of an op-amp.
Input offset voltage—The equivalent DC voltage that must be applied to one of the input terminals of an op-amp to produce a zero output voltage with the other input grounded.
Input resistance—The internal input impedance of an op-amp as seen at either the inverting or the noninverting input to ground while the remaining input terminal is grounded.
Instrumentation amplifier—A precision, high-impedance difference amplifier whose gain can be varied by a single resistor.
Inverting amplifier—An op-amp circuit whose output is 180° out of phase with the amplifier's input.
Inverting comparator—A circuit whose output goes negative when the input signal is more positive than the reference voltage, and goes positive when the input signal is less than the reference voltage.
Logarithmic amplifier—A nonlinear circuit whose output signal is proportional to the logarithm of its input signal.
Low-pass filter—A filter network that allows signals below a certain frequency to pass through unaffected, but rejects signal frequencies above this frequency.
Multiple feedback filter—A type of op-amp active filter having multiple feedback paths connected between the output and the op-amp's inverting input such that the passband output signal is 180° out of phase with the input.
Noninverting amplifier—An op-amp circuit whose output is in phase with the amplifier's input.
Noninverting comparator—A circuit whose output goes positive when the input signal is more positive than the reference voltage, and goes negative when the input signal is less than the reference voltage.
Norton amplifier—An integrated-circuit device that amplifies the difference in the currents flowing into its input terminals. Also called a *current differencing amplifier (CDA)*.
Notch filter—A filter network that rejects signals between two frequencies, but allows signals at frequencies above and below this range to pass through unaffected. Also called a *band-stop* or *band-reject filter*.
Open-loop gain—The voltage gain of the op-amp without any external feedback.
Open-loop mode—Referring to when an op-amp is operated without external feedback.
Order—An integer value that relates the rolloff or selectivity of a low-pass or high-pass filter.
Output resistance—The internal output resistance as seen from the op-amp's output terminal to ground.
Output voltage swing—The maximum peak output voltage that the op-amp can produce without saturation or clipping, and is typically about two diode voltage drops less than the corresponding supply voltages.
Passband—The filter's frequency range where the input signal passes through with no attenuation.
Peak detector— A diode circuit with an op-amp that captures the peak positive or negative excursion of an input signal for an infinite period of time until it is reset to zero.
Phase-shift oscillator—A low-frequency sine-wave oscillator circuit that uses three identical RC networks as the feedback element to produce a 180° phase shift at the desired frequency of oscillation.
Precision rectifier—A rectifier circuit which is either half- or full-wave built around an op-amp.
Rolloff—The rate at which the stopband response of a filter network changes with frequency, expressed in units of dB/octave or dB/decade.

Sallen-Key filter—A type of low- or high-pass active filter network whose passband output is in phase with the input signal.

Sample-hold amplifier (S/H)— A circuit that captures and holds the value of a rapidly changing input voltage at a specific point in time.

Schmitt trigger—A comparator with hysteresis, triggered at predetermined voltage levels.

Slew rate—The time rate of change of the output voltage with the op-amp circuit having a unity closed-loop gain.

State-variable filter—A universal active filter network using three op-amps to achieve simultaneous low-pass, high-pass, and bandpass responses.

Stopband—The filter's frequency range where the input signal is attenuated.

Summing amplifier—An op-amp circuit whose output is proportional to the algebraic sum of its inputs.

Twin-T oscillator—A low-frequency oscillator circuit that uses a pair of RC networks (each arranged in a *T* fashion) and is connected in parallel as the feedback element so that its impedance is zero and phase shift is 180° at the desired frequency of oscillation. Also called *parallel-T oscillator*.

Unity-gain frequency—The frequency where the op-amp's open-loop gain is unity. It is also called either the *unity-gain crossover frequency* or the *small signal unity-gain bandwidth*.

Voltage follower—An op-amp circuit whose output is exactly the same as its input. It is frequently called a buffer as it isolates a load from a given signal source.

Wien-bridge oscillator—An RC-oscillator circuit that uses a an RC lead-lag network as the feedback elements to achieve oscillation.

Window comparator—A circuit that detects the presence of a voltage between two specified voltage limits. Also called a *double-ended comparator*.

Zero crossing detector— A comparator circuit whose output changes when the input signal crosses zero volts. Also called a *polarity detector*.

Appendix A

Data Sheets

This appendix contains the following data sheets:

1. CA3140, CMOS (RCA)
2. μA747, dual general purpose (Texas Instruments)
3. LF351, JFET-input (Texas Instruments)
4. LM101/201/301, high-performance (Texas Instruments)
5. LM218/318, high-performance (Texas Instruments)
6. LM2900/3900, quad Norton (Texas Instruments)

The data sheet for the μA741 general-purpose operational amplifier (Signetics) is presented in Figure 2-1.

CA3140, CA3140A, CA3140B Types
BiMOS Operational Amplifiers
With MOS/FET Input, Bipolar Output

The CA3140B, CA3140A, and CA3140 are integrated-circuit operational amplifiers that combine the advantages of high-voltage PMOS transistors with high-voltage bipolar transistors on a single monolithic chip. Because of this unique combination of technologies, this device can now provide designers, for the first time, with the special performance features of the CA3130 COS/MOS operational amplifiers and the versatility of the 741 series of industry-standard operational amplifiers.

The CA3140, CA3140A, and CA3140 BiMOS operational amplifiers feature gate-protected MOS/FET (PMOS) transistors in the input circuit to provide very-high-input impedance, very-low-input current, and high-speed performance. The CA3140B operates at supply voltages from 4 to 44 volts; the CA3140A and CA3140 from 4 to 36 volts (either single or dual supply). These operational amplifiers are internally phase-compensated to achieve stable operation in unity-gain follower operation, and, additionally, have access terminals for a supplementary external capacitor if additional frequency roll-off is desired. Terminals are also provided for use in applications requiring input offset-voltage nulling. The use of PMOS field-effect transistors in the input stage results in common-mode input-voltage capability down to 0.5 volt below the negative-supply terminal, an important attribute for single-supply applications. The output stage uses bipolar transistors and includes built-in protection against damage from load-terminal short-circuiting to either supply-rail or to ground.

The CA3140 Series has the same 8-lead terminal pin-out used for the "741" and other industry-standard operational amplifiers. They are supplied in either the standard 8-lead TO-5 style package (T suffix), or in the 8-lead dual-in-line formed-lead TO-5 style package "DIL-CAN" (S suffix). The CA3140 is available in chip form (H suffix). The CA3140A and CA3140 are also available in an 8-lead dual-in-line plastic package (Mini-DIP-E suffix). The CA3140B is intended for operation at supply voltages ranging from 4 to 44 volts, for applications requiring premium-grade specifications and with electrical limits established for operations over the range from -55°C to $+125^\circ$C. The CA3140A and CA3140 are for operation at supply voltages up to 36 volts (± 18 volts). The CA3140 ages up to 36 volts (± 18 volts). All types can be operated safely over the temperature range from -55°C to $+125^\circ$C.

Features:
- MOS/FET Input Stage
  - (a) Very high input impedance (Z_{IN}) – 1.5 TΩ typ.
  - (b) Very low input current (I_I) – 10 pA typ. at ± 15 V
  - (c) Low input-offset voltage (V_{IO}) – to 2 mV max.
  - (d) Wide common-mode input-voltage range (V_{ICR}) – can be swung 0.5 volt below negative supply-voltage rail
  - (e) Output swing complements input common-mode range
  - (f) Rugged input stage – bipolar diode protected
- Directly replaces industry type 741 in most applications
- Includes numerous industry operational amplifier categories such as general-purpose, FET input, wideband (high slew rate)
- Operation from 4-to-44 volts
  Single or Dual supplies
- Internally compensated
- Characterized for ± 15-volt operation and for TTL supply systems with operation down to 4 volts
- Wide bandwidth – 4.5 MHz unity gain at ± 15 V or 30 V; 3.7 MHz at 5 V
- High voltage-follower slew rate – 9 V/μs
- Fast settling time – 1.4 μs typ. to 10 mV with a 10-V$_{p-p}$ signal
- Output swings to within 0.2 volt of negative supply
- Strobable output stage

MAXIMUM RATINGS, Absolute-Maximum Values:

	CA3140, CA3140A	CA3140B
DC SUPPLY VOLTAGE (BETWEEN V+ AND V– TERMINALS)	36 V	44 V
DIFFERENTIAL-MODE INPUT VOLTAGE	± 8 V	± 8 V
COMMON-MODE DC INPUT VOLTAGE	(V+ +8 V) to (V– –0.5 V)	
INPUT-TERMINAL CURRENT	1 mA	
DEVICE DISSIPATION:		
WITHOUT HEAT SINK –		
UP TO 55°C	630 mW	
ABOVE 55°C	Derate linearly 6.67 mW/°C	
WITH HEAT SINK –		
Up to 55°C	1 W	
Above 55°C	Derate linearly 16.7 mW/°C	
TEMPERATURE RANGE:		
OPERATING (ALL TYPES)	–55 to +125°C	
STORAGE (ALL TYPES)	–65 to +150°C	
OUTPUT SHORT-CIRCUIT DURATION*	INDEFINITE	
LEAD TEMPERATURE (DURING SOLDERING):		
AT DISTANCE 1/16 $\pm$ 1/32 INCH (1.59 $\pm$ 0.79 MM) FROM CASE FOR 10 SECONDS MAX.	+265°C	

* Short circuit may be applied to ground or to either supply.

Applications:
- Ground-referenced single-supply amplifiers in automobile and portable instrumentation
- Sample and hold amplifiers
- Long-duration timers/multivibrators (microseconds—minutes—hours)
- Photocurrent instrumentation
- Peak detectors   ■ Active filters
- Comparators
- Interface in 5 V TTL systems & other low-supply voltage systems
- All standard operational amplifier applications
- Function generators   ■ Tone controls
- Power supplies   ■ Portable instruments
- Intrusion alarm systems

Fig. 1 – Functional diagrams of the CA3140 series.

A-1. CA3140, CMOS (RCA)

CA3140, CA3140A, CA3140B Types

TYPICAL ELECTRICAL CHARACTERISTICS

CHARACTERISTIC		TEST CONDITIONS $V^+ = +15$ V $V^- = -15$ V $T_A = 25°C$	CA3140B (T,S)	CA3140A (T,S,E)	CA3140 (T,S,E)	UNITS
Input Offset Voltage Adjustment Resistor		Typ. Value of Resistor Between Term. 4 and 5 or 4 and 1 to Adjust Max. V_{IO}	43	18	4.7	kΩ
Input Resistance	R_I		1.5	1.5	1.5	TΩ
Input Capacitance	C_I		4	4	4	pF
Output Resistance	R_O		60	60	60	Ω
Equivalent Wideband Input Noise Voltage (See Fig. 39)	e_n	BW = 140 kHz R_S = 1 MΩ	48	48	48	μV
Equivalent Input Noise Voltage (See Fig.10)	e_n	f = 1 kHz R_S =	40	40	40	nV/$\sqrt{Hz}$
		f = 10 kHz 100 Ω	12	12	12	
Short-Circuit Current to Opposite Supply	Source I_{OM}^+		40	40	40	mA
	Sink I_{OM}^-		18	18	18	mA
Gain-Bandwidth Product, (See Figs. 5 &18)	f_T		4.5	4.5	4.5	MHz
Slew Rate, (See Fig.6)	SR		9	9	9	V/μs
Sink Current From Terminal 8 To Terminal 4 to Swing Output Low			220	220	220	μA
Transient Response: Rise Time Overshoot (See Fig. 37)	t_r	R_L = 2 kΩ C_L = 100 pF	0.08 10	0.08 10	0.08 10	μs %
Settling Time at 10 V_{p-p} (See Fig.17)	1 mV t_s 10 mV	R_L = 2 kΩ C_L = 100 pF Voltage Follower	4.5 1.4	4.5 1.4	4.5 1.4	μs

CIRCUIT DESCRIPTION

Fig.2 is a block diagram of the CA3140 Series PMOS Operational Amplifiers. The input terminals may be operated down to 0.5 V below the negative supply rail. Two class A amplifier stages provide the voltage gain, and a unique class AB amplifier stage provides the current gain necessary to drive low-impedance loads.

A biasing circuit provides control of cascoded constant-current flow circuits in the first and second stages. The CA3140 includes an on-chip phase-compensating capacitor that is sufficient for the unity gain voltage-follower configuration.

Input Stages — The schematic circuit diagram of the CA3140 is shown in Fig.3. It consists of a differential-input stage using PMOS field-effect transistors (Q9, Q10) working into a mirror pair of bipolar transistors (Q11, Q12) functioning as load resistors together with resistors R2 through R5. The mirror-pair transistors also function as a differential-to-single-ended converter to provide base-current drive to the second-stage bipolar transistor (Q13). Offset nulling, when desired, can be effected with a 10-kΩ potentiometer connected across terminals 1 and 5 and with its slider arm connected to terminal 4. Cascode-connected bipolar transistors Q2, Q5 are the constant-current source for the input stage. The base-biasing circuit for the constant-current source is described subsequently. The small diodes D3, D4, D5 provide gate-oxide protection against high-voltage transients, e.g., static electricity.

Second Stage — Most of the voltage gain in the CA3140 is provided by the second amplifier stage, consisting of bipolar transistor Q13 and its cascode-connected load resistance provided by bipolar transistors Q3, Q4. On-chip phase compensation, sufficient for a majority of the applications is provided by C1. Additional Miller-Effect compensation (roll-off) can be accomplished, when desired, by simply connecting a small capacitor between terminals 1 and 8. Terminal 8 is also used to strobe the output stage into quiescence. When terminal 8 is tied to the negative supply rail (terminal 4) by mechanical or electrical means, the output terminal 6 swings low, i.e., approximately to terminal 4 potential.

Output Stage — The CA3140 Series circuits employ a broadband output stage that can sink loads to the negative supply to complement the capability of the PMOS input stage when operating near the negative rail. Quiescent current in the emitter-follower cascode circuit (Q17, Q18) is established by transistors (Q14, Q15) whose base-currents are "mirrored" to current flowing through diode D2 in the bias circuit section. When the CA3140 is operating such that output terminal 6 is sourcing current, transistor Q18 functions as an emitter-follower to source current from the V+ bus (terminal 7), via D7, R9, and R11. Under these conditions, the collector potential of Q13 is sufficiently high to permit the necessary flow of base current to emitter follower Q17 which, in turn, drives Q18.

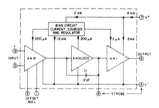

Fig.2 — Block diagram of CA3140 series.

A-1. CA3140, CMOS (RCA)

241

CA3140, CA3140A, CA3140B Types

ELECTRICAL CHARACTERISTICS FOR EQUIPMENT DESIGN
At $V^+ = 15$ V, $V^- = 15$ V, $T_A = 25°C$ Unless Otherwise Specified

CHARACTERISTIC	LIMITS									UNITS
	CA3140B			CA3140A			CA3140			
	Min.	Typ.	Max.	Min.	Typ.	Max.	Min.	Typ.	Max.	
Input Offset Voltage, $\|V_{IO}\|$	–	0.8	2	–	2	5	–	5	15	mV
Input Offset Current, $\|I_{IO}\|$	–	0.5	10	–	0.5	20	–	0.5	30	pA
Input Current, I_I	–	10	30	–	10	40	–	10	50	pA
Large-Signal Voltage Gain, A_{OL}● (See Figs. 4,18)	50 k	100 k	–	20 k	100 k	–	20 k	100 k	–	V/V
	94	100	–	86	100	–	86	100	–	dB
Common-Mode Rejection Ratio, CMRR (See Fig.9)	–	20	50	–	32	320	–	32	320	µV/V
	86	94	–	70	90	–	70	90	–	dB
Common-Mode Input-Voltage Range, V_{ICR} (See Fig.20)	–15	–15.5 to +12.5	12	–15	–15.5 to +12.5	12	–15	–15.5 to +12.5	1.1	V
Power-Supply Rejection $\Delta V_{IO}/\Delta V$ Ratio, PSRR (See Fig.11)	–	32	100	–	100	150	–	100	150	µV/V
	80	90	–	76	80	–	76	80	–	dB
Max. Output Voltage■ V_{OM}^+ (See Figs.13,20) V_{OM}^-	+12	13	–	+12	13	–	+12	13	–	V
	–14	–14.4	–	–14	–14.4	–	–14	–14.4	–	
Supply Current, I^+ (See Fig.7)	–	4	6	–	4	6	–	4	6	mA
Device Dissipation, P_D	–	120	180	–	120	180	–	120	180	mW
Input Current, I_I▲ (See Fig.19)	–	10	30	–	10	–	–	10	–	nA
Input Offset Voltage $\|V_{IO}\|$▲	–	1.3	3	–	3	–	–	10	–	mV
Input Offset Voltage Temp. Drift, $\Delta V_{IO}/\Delta T$	–	5	–	–	6	–	–	8	–	µV/°C
Large-Signal Voltage Gain, A_{OL}▲ (See Figs.4,18)	20 k	100 k	–	–	100 k	–	–	100 k	–	V/V
	86	100	–	–	100	–	–	100	–	dB
Max. Output Voltage,* V_{OM}^+ V_{OM}^-	+19	+19.5	–	–	–	–	–	–	–	V
	–21	–21.4	–	–	–	–	–	–	–	
Large-Signal Voltage Gain, A_{OL}♦*	20 k	50 k	–	–	–	–	–	–	–	V/V
	86	94	–	–	–	–	–	–	–	dB

● At $V_O = 26V_{p-p}$, +12V, –14V and $R_L = 2$ kΩ.   ■ At $R_L = 2$ kΩ.
▲ At $T_A = -55°C$ to $+125°C$, $V^+ = 15$ V, $V^- = 15$ V, $V_O = 26V_{p-p}$, $R_L = 2$ kΩ.
♦ At $V_O = +19$ V, –21 V, and $R_L = 2$ kΩ.   * At $V^+ = 22$ V, $V^- = 22$ V.

When the CA3140 is operating such that output terminal 6 is sinking current to the V– bus, transistor Q16 is the current-sinking element. Transistor Q16 is mirror-connected to D6, R7, with current fed by way of Q21, R12, and Q20. Transistor Q20, in turn, is biased by current-flow through R13, zener D8, and R14. The dynamic current-sink is controlled by voltage-level sensing. For purposes of explanation, it is assumed that output terminal 6 is quiescently established at the potential mid-point between the V+ and V– supply rails. When output-current sinking-mode operation is required, the collector potential of transistor Q13 is driven below its quiescent level, thereby causing Q17, Q18 to decrease the output voltage at terminal 6. Thus, the gate terminal of PMOS transistor Q21 is displaced toward the V– bus, thereby reducing the channel resistance of Q21. As a consequence, there is an incremental increase in current flow through Q20, R12, Q21, D6, R7, and the base of Q16. As a result, Q16 sinks current from terminal 6 in direct response to the incremental change in output voltage caused by Q18. This sink current flows regardless of load; any excess current is internally supplied by the emitter-follower Q18. Short-circuit protection of the output circuit is provided by Q19, which is driven into conduction by the high voltage drop developed across R11 under output short-circuit conditions. Under these conditions, the collector of Q19 diverts current from Q4 so as to reduce the base-current drive from Q17, thereby limiting current flow in Q18 to the short-circuited load terminal.

Bias Circuit – Quiescent current in all stages (except the dynamic current sink) of the CA3140 is dependent upon bias current flow in R1. The function of the bias circuit is to establish and maintain constant-current flow through D1, Q6, Q8 and D2. D1 is a diode-connected transistor mirror-connected in parallel with the base-emitter junctions of Q1, Q2, and Q3. D1 may be considered as a current-sampling diode that senses the emitter current of Q6 and automatically adjusts the base current of Q6 (via Q1) to maintain a constant current through Q6, Q8, D2. The base-currents in Q2, Q3 are also determined by constant-current flow D1. Furthermore, current in diode-connected transistor D2 establishes the currents in transistors Q14 and Q15.

A-1. CA3140, CMOS (RCA)

CA3140, CA3140A, CA3140B Types

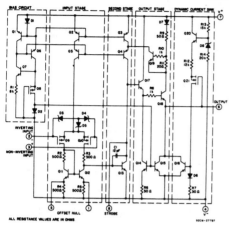

Fig.3 – Schematic diagram of CA3140 series.

TYPICAL ELECTRICAL CHARACTERISTICS FOR DESIGN GUIDANCE
At $V^+ = 5$ V, $V^- = 0$ V, $T_A = 25°C$

CHARACTERISTIC		CA3140B (T,S)	CA3140A (T,S,E)	CA3140 (T,S,E)	UNITS
Input Offset Voltage	$|V_{IO}|$	0.8	2	5	mV
Input Offset Current	$|I_{IO}|$	0.1	0.1	0.1	pA
Input Current	I_I	2	2	2	pA
Input Resistance		1	1	1	TΩ
Large-Signal Voltage Gain	A_{OL}	100 k	100 k	100 k	V/V
(See Figs.4,18)		100	100	100	dB
Common-Mode Rejection Ratio,	CMRR	20	32	32	µV/V
		94	90	90	dB
Common-Mode Input-Voltage Range	V_{ICR}	–0.5	–0.5	–0.5	V
(See Fig.20)		2.6	2.6	2.6	
Power-Supply Rejection Ratio	$\Delta V_{IO}/\Delta V^+$	32	100	100	µV/V
		90	80	80	dB
Maximum Output Voltage	V_{OM}^+	3	3	3	V
(See Figs.13,20)	V_{OM}^-	0.13	0.13	0.13	
Maximum Output Current:					
Source	I_{OM}^+	10	10	10	mA
Sink	I_{OM}^-	1	1	1	
Slew Rate (See Fig.6)		7	7	7	V/µs
Gain-Bandwidth Product (See Fig.5)	f_T	3.7	3.7	3.7	MHz
Supply Current (See Fig.7)	I^+	1.6	1.6	1.6	mA
Device Dissipation	P_D	8	8	8	mW
Sink Current from Term. 8 to Term. 4 to Swing Output Low		200	200	200	µA

Fig.4 – Open-loop voltage gain vs supply voltage and temperature.

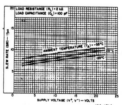

Fig.5 – Gain-bandwidth product vs supply voltage and temperature.

Fig.6 – Slew rate vs supply voltage and temperature.

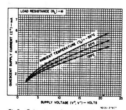

Fig.7 – Quiescent supply current vs supply voltage and temperature.

A-1. CA3140, CMOS (RCA)

243

CA3140, CA3140A, CA3140B Types

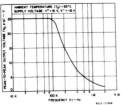

Fig.8 — *Maximum output voltage swing vs frequency.*

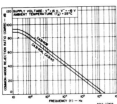

Fig.9 — *Common-mode rejection ratio vs frequency.*

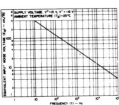

Fig.10 — *Equivalent input noise voltage vs frequency.*

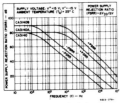

Fig.11 — *Power supply rejection ratio vs frequency.*

APPLICATIONS CONSIDERATIONS

Wide dynamic range of input and output characteristics with the most desirable high input-impedance characteristic is achieved in the CA3140 by the use of an unique design based upon the PMOS-Bipolar process. Input-common-mode voltage range and output-swing capabilities are complementary, allowing operation with the single supply down to four volts.

The wide dynamic range of these parameters also means that this device is suitable for many single-supply applications, such as, for example, where one input is driven below the potential of terminal 4 and the phase sense of the output signal must be maintained — a most important consideration in comparator applications.

OUTPUT CIRCUIT CONSIDERATIONS

Excellent interfacing with TTL circuitry is easily achieved with a single 6.2-volt zener diode connected to terminal 8 as shown in Fig.12. This connection assures that the maximum output signal swing will not go more positive than the zener voltage minus two base-to-emitter voltage drops within the CA3140. These voltages are independent of the operating supply voltage.

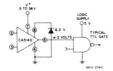

Fig.12 — *Zener clamping diode connected to terminals 8 and 4 to limit CA3140 output swing to TTL levels.*

Fig.13 shows output current-sinking capabilities of the CA3140 at various supply voltages. Output voltage swing to the negative supply rail permits this device to operate both power transistors and thyristors directly without the need for level-shifting circuitry usually associated with the 741 series of operational amplifiers.

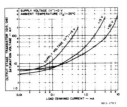

Fig.13 — *Voltage across output transistors Q15 and Q16 vs load current.*

Fig.16 show some typical configurations. Note that a series resistor, R_L, is used in both cases to limit the drive available to the driven device. Moreover, it is recommended that a series diode and shunt diode be used at the thyristor input to prevent large negative transient surges that can appear at the gate of thyristors, from damaging the integrated circuit.

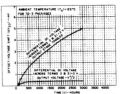

Fig.14 — *Typical incremental offset-voltage shift vs operating life.*

OFFSET-VOLTAGE NULLING

The input-offset voltage can be nulled by connecting a 10-kΩ potentiometer between terminals 1 and 5 and returning its wiper arm to terminal 4, see Fig.15a. This technique, however, gives more adjustment range than required and therefore, a considerable portion of the potentiometer rotation is not fully utilized. Typical values of series resistors that may be placed at either end of the potentiometer, see Fig.15b, to optimize its utilization range are given in the table "Typical Electrical Characteristics" shown in this bulletin.

An alternate system is shown in Fig.15c. This circuit uses only one additional resistor of approximately the value shown in the table. For potentiometers, in which the resistance does not drop to zero ohms at either end of rotation, a value of resistance 10% lower than the values shown in the table should be used.

LOW-VOLTAGE OPERATION

Operation at total supply voltages as low as 4 volts is possible with the CA3140. A current regulator based upon the PMOS threshold voltage maintains reasonable constant operating current and hence consistent performance down to these lower voltages.

The low-voltage limitation occurs when the upper extreme of the input common-mode voltage range extends down to the voltage at terminal 4. This limit is reached at a total supply voltage just below 4 volts. The output voltage range also begins to extend down to the negative supply rail, but is slightly higher than that of the input. Fig.20 shows these characteristics and shows that with 2-volt dual supplies, the lower extreme of the input common-mode voltage range is below ground potential.

A-1. CA3140, CMOS (RCA)

CA3140, CA3140A, CA3140B Types

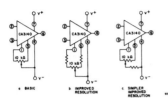

Fig. 15 — Three offset-voltage nulling methods.

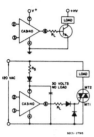

Fig. 16 — Methods of utilizing the $V_{CE(sat)}$ sinking-current capability of the CA3140 series.

BANDWIDTH AND SLEW RATE

For those cases where bandwidth reduction is desired, for example, broadband noise reduction, an external capacitor connected between terminals 1 and 8 can reduce the open-loop −3 dB bandwidth. The slew rate will, however, also be proportionally reduced by using this additional capacitor. Thus, a 20% reduction in bandwidth by this technique will also reduce the slew rate by about 20%.

Fig. 17 shows the typical settling time required to reach 1 mV or 10 mV of the final value for various levels of large signal inputs for the voltage-follower and inverting unity-gain amplifiers. The exceptionally fast settling time characteristics shown in Fig. 18 are largely due to the high combination of high gain and wide bandwidth of the CA3140.

INPUT CIRCUIT CONSIDERATIONS

As mentioned previously, the amplifier inputs can be driven below the terminal 4 potential, but a series current-limiting resistor is recommended to limit the maximum input terminal current to less than 1 mA to prevent damage to the input protection circuitry.

Moreover, some current-limiting resistance should be provided between the inverting input and the output when the CA3140 is used as a unity-gain voltage follower. This resistance prevents the possibility of ex-

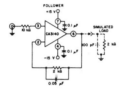

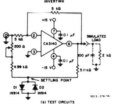

Fig. 17 — Input voltage vs settling time.

tremely large input-signal transients from forcing a signal through the input-protection network and directly driving the internal constant-current source which could result in positive feedback via the output terminal. A 3.9-kΩ resistor is sufficient.

The typical input current is in the order of 10 pA when the inputs are centered at nominal device dissipation. As the output supplies load current, device dissipation will increase, raising the chip temperature and resulting in increased input current. Fig. 19 shows typical input-terminal current versus ambient temperature for the CA3140.

It is well known that MOS/FET devices can exhibit slight changes in characteristics (for example, small changes in input offset voltage) due to the application of large differential input voltages that are sustained over long periods at elevated temperatures. Both applied voltage and temperature accelerate these changes. The process is reversible and offset voltage shifts of the opposite polarity reverse the offset. Fig. 14 shows the typical offset voltage change as a function of various stress voltages at the maximum rating of 125°C (for TO-5); at lower temperatures (TO-5 and plastic), for example, at 85°C, this change in voltage is considerably less. In typical linear applications, where the differential voltage is small and symmetrical, these incremental changes are of about the same magnitude as those encountered in an operational amplifier employing a bipolar a transistor input stage.

SUPER SWEEP FUNCTION GENERATOR

A function generator having a wide tuning range is shown in Fig. 21. The 1,000,000/1 adjustment range is accomplished by a single variable potentiometer or by an auxiliary sweeping signal. The CA3140 functions as a non-inverting read-out amplifier of the tri-

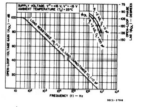

Fig. 18 — Open-loop voltage gain and phase lag vs frequency.

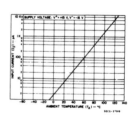

Fig. 19 — Input current vs ambient temperature.

A-1. CA3140, CMOS (RCA)

245

CA3140, CA3140A, CA3140B Types

angular signal developed across the integrating capacitor network connected to the output of the CA3080A current source.

Buffered triangular output signals are then applied to a second CA3080 functioning as a high-speed hysteresis switch. Output from the switch is returned directly back to the

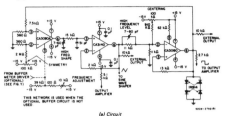

(a) Circuit

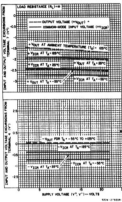

Fig.20 — Output-voltage-swing capability and common-mode input-voltage range vs supply voltage and temperature.

(b1) Function generator sweeping

(b2) Function generator with fixed frequencies

Fig.21 — Function generator.

input of the CA3080A current source, thereby, completing the positive feedback loop.

The triangular output level is determined by the four 1N914 level-limiting diodes of the second CA3080 and the resistor-divider network connected to terminal No.2 (input) of the CA3080. These diodes establish the input trip level to this switching stage and, therefore, indirectly determine the amplitude of the output triangle.

Compensation for propagation delays around the entire loop is provided by one adjustment on the input of the CA3080. This adjustment, which provides for a constant generator amplitude output, is most easily made while the generator is sweeping. High-frequency ramp linearity is adjusted by the single 7-to-60 pF capacitor in the output of the CA3080A.

It must be emphasized that only the CA-3080A is characterized for maximum output linearity in the current-generator function.

METER DRIVER AND BUFFER AMPLIFIER

Fig. 22 shows the CA3140 connected as a meter driver and buffer amplifier. Low driving impedance is required of the CA-3080A current source to assure smooth operation of the Frequency Adjustment Control. This low-driving impedance requirement is easily met by using a CA3140 connected as a voltage follower. Moreover, a meter may be placed across the input to the CA3080A to give a logarithmic analog indication of the function generators frequency.

Analog frequency readout is readily accomplished by the means described above because the output current of the CA3080A varies approximately one decade for each 60-mV change in the applied voltage, V_{ABC} (voltage between terminals 5 and 4 of the CA3080A of the function generator). Therefore, six decades represent 360-mV change in V_{ABC}.

Now, only the reference voltage must be established to set the lower limit on the meter. The three remaining transistors from

(c) Interconnections

1V/DIV and 1 sec/DIV

Three tone test signals, highest frequency ≥ 0.5 MHz. Note the slight assymmetry at the three-second/cycle signal. This assymmetry is due to slightly different positive and negative integration from the CA3080A and from the pc board and component leakages at the 100-pA level.

the CA3086 Array used in the sweep generator are used for this reference voltage. In addition, this reference generator arrangement tends to track ambient temperature variations, and thus compensates for the effects of the normal negative temperature coefficient of the CA3080A V_{ABC} terminal voltage.

Another output voltage from the reference generator is used to insure temperature tracking of the lower end of the Frequency Adjustment Potentiometer. A large series resistance simulates a current source, assuring similar temperature coefficients at both ends of the Frequency Adjustment Control.

To calibrate this circuit, set the Frequency Adjustment Potentiometer at its low end. Then adjust the Minimum Frequency Calibration Control for the lowest frequency. To

A-1. CA3140, CMOS (RCA)

CA3140, CA3140A, CA3140B Types

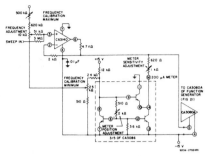

Fig. 22 — Meter driver and buffer amplifier.

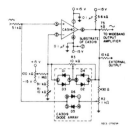

Fig. 23 — Sine-wave shaper.

potentiometer connected between terminals 2 and 6 of the CA3140 and the 9.1-kΩ resistor and 10-kΩ potentiometer from terminal 2 to ground. Two break points are established by diodes D_1 through D_4. Positive feedback via D_5 and D_6 establishes the zero slope at the maximum and minimum levels of the sine wave. This technique is necessary because the voltage-follower configuration approaches unity gain rather than the zero gain required to shape the sine wave at the two extremes.

This circuit can be adjusted most easily with a distortion analyzer, but a good first approximation can be made by comparing the output signal with that of a sine-wave generator. The initial slope is adjusted with the potentiometer R_1, followed by an adjustment of R_2. The final slope is established by adjusting R_3, thereby adding additional segments that are contributed by these diodes. Because there is some interaction among these controls, repetition of the adjustment procedure may be necessary

SWEEPING GENERATOR

Fig. 24 shows a sweeping generator. Three CA3140's are used in this circuit. One CA3140 is used as an integrator, a second device is used as a hysteresis switch that determines the starting and stopping points of the sweep. A third CA3140 is used as a logarithmic shaping network for the log function. Rates and slopes, as well as sawtooth, triangle, and logarithmic sweeps are generated by this circuit.

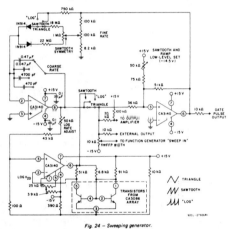

Fig. 24 — Sweeping generator.

establish the upper frequency limit, set the Frequency Adjustment Potentiometer to its upper end and then adjust the Maximum Frequency Calibration Control for the maximum frequency. Because there is interaction among these controls, repetition of the adjustment procedure may be necessary.

Two adjustments are used for the meter. The meter sensitivity control sets the meter-scale width of each decade, while the meter position control adjusts the pointer on the scale with negligible effect on the sensitivity adjustment. Thus, the meter sensitivity adjustment control calibrates the meter so that it deflects 1.6 of full scale for each decade change in frequency.

SINE-WAVE SHAPER

The circuit shown in Fig. 23 uses a CA3140 as a voltage follower in combination with diodes from the CA3019 Array to convert the triangular signal from the function generator to a sine-wave output signal having typically less than 2% THD. The basic zero-crossing slope is established by the 10-kΩ

WIDEBAND OUTPUT AMPLIFIER

Fig. 25 shows a high-slew-rate, wideband amplifier suitable for use as a 50-ohm transmission-line driver. This circuit, when used in conjunction with the function generator and sine-wave shaper circuits shown in Figs. 21 and 23 provides 18 volts peak-to-peak output open-circuited, or 9 volts peak-to-peak output when terminated in 50 ohms. The slew rate required of this amplifier is 28 volts/μs (18 volts peak-to-peak x π x 0.5 MHz).

A-1. CA3140, CMOS (RCA)

CA3140, CA3140A, CA3140B Types

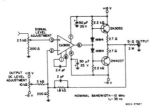

Fig. 25 — Wideband output amplifier.

POWER SUPPLIES

High input-impedance, common-mode capability down to the negative supply and high output-drive current capability are key factors in the design of wide-range output-voltage supplies that use a single input voltage to provide a regulated output voltage that can be adjusted from essentially 0 to 24 volts.

Unlike many regulator systems using comparators having a bipolar transistor-input stage, a high-impedance reference-voltage divider from a single supply can be used in connection with the CA3140 (see Fig. 26).

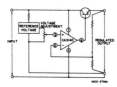

Fig. 26 — Basic single-supply voltage regulator showing voltage-follower configuration.

Essentially, the regulators, shown in Figs. 27 and 28, are connected as non-inverting power operational amplifiers with a gain of 3.2. An 8-volt reference input yields a maximum output voltage slightly greater than 25 volts. As a voltage follower, when the reference input goes to 0 volts the output will be 0 volts. Because the offset voltage is also multiplied by the 3.2 gain factor, a potentiometer is needed to null the offset voltage.

Series pass transistors with high I_{CBO} levels will also prevent the output voltage from reaching zero because there is a finite voltage drop (V_{CE}sat) across the output of the CA3140 (see Fig.13). This saturation voltage level may indeed set the lowest voltage obtainable.

The high impedance presented by terminal 8 is advantageous in effecting current limiting. Thus, only a small signal transistor is

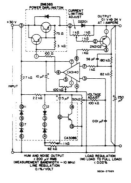

Fig. 27 — Regulated power supply.

required for the current-limit sensing amplifier. Resistive decoupling is provided for this transistor to minimize damage to it or the CA3140 in the event of unusual input or output transients on the supply-rail.

Figs. 27 and 28, show circuits in which a D2201 high-speed diode is used for the current sensor. This diode was chosen for its slightly higher forward-voltage drop characteristic thus giving greater sensitivity. It must be emphasized that heat sinking of this diode is essential to minimize variation of the current trip point due to internal heating of the diode. That is, 1 ampere at 1 volt forward drop represents one watt which can result in significant regenerative changes in the current trip point as the diode temperature rises. Placing the small-signal reference amplifier in the proximity of the current-sensing diode also helps minimize the variability in the trip level due to the negative temperature coefficient of the diode.

In spite of those limitations, the current limiting point can easily be adjusted over the range from 10 mA to 1 ampere with a single adjustment potentiometer. If the temperature stability of the current-limiting

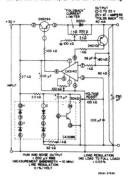

Fig. 28 — Regulated power supply with "foldback" current limiting.

(a)
SUPPLY TURN-ON AND TURN-OFF CHARACTERISTICS
(5 VOLTS/DIV AND 1 s/DIV)

(b)
TRANSIENT RESPONSE
TOP TRACE: OUTPUT VOLTAGE
(200 mV/DIV AND 5μs/DIV)
BOTTOM TRACE: COLLECTOR OF LOAD SWITCHING TRANSISTOR,
LOAD = 1 AMPERE
(5 VOLTS/DIV AND 5μs/DIV)

Fig. 29 — Waveforms of dynamic characteristics of power supply currents shown in Figs. 29 and 30.

A-1. CA3140, CMOS (RCA)

CA3140, CA3140A, CA3140B Types

system is a serious consideration, the more usual current-sampling resistor-type of circuitry should be employed.

A power Darlington transistor (in a heat sink TO-3 case), is used as the series-pass element for the conventional current-limiting system, Fig. 27, because high-power Darlington dissipation will be encountered at low output voltage and high currents.

A small heat-sink VERSAWATT transistor is used as the series-pass element in the foldback current system, Fig.28, since dissipation levels will only approach 10 watts. In this system, the D2201 diode is used for current sampling. Foldback is provided by the 3 kΩ and 100 kΩ divider network connected to the base of the current-sensing transistor.

Both regulators, Figs. 27 and 28, provide better than 0.02% load regulation. Because there is constant loop gain at all voltage settings, the regulation also remains constant. Line regulation is 0.1% per volt. Hum and noise voltage is less than 200 μV as read with a meter having a 10-MHz bandwidth. Fig.31 (a) shows the turn ON and turn OFF characteristics of both regulators. The slow turn-on rise is due to the slow rate of rise of the reference voltage. Fig. 29 (b) shows the transient response of the regulator with the switching of a 20-Ω load at 20 volts output.

TONE CONTROL CIRCUITS

High-slew-rate, wide-bandwidth, high-output voltage capability and high input impedance are all characteristics required of tone-control amplifiers. Two tone control circuits that exploit these characteristics of the CA3140 are shown in Figs. 30 and 31. The first circuit, shown in Fig. 31, is the Baxandall tone-control circuit which provides unity gain at midband and uses standard linear potentiometers. The high input impedance of the CA3140 makes possible the use of low-cost, low-value, small-size capacitors, as well as reduced load of the driving stage.

Bass treble boost and cut are ± 15 dB at 100 Hz and 10 kHz, respectively. Full peak-to-peak output is available up to at least 20 kHz due to the high slew rate of the CA3140. The amplifier gain is –3 dB down from its "flat" position at 70 kHz.

Fig. 30 shows another tone-control circuit with similar boost and cut specifications. The wideband gain of this circuit is equal to the ultimate boost or cut plus one, which in this case is a gain of eleven. For 20-dB boost and cut, the input loading of this circuit is essentially equal to the value of the resistance from terminal No.3 to ground. A detailed analysis of this circuit is given in "An IC Operational Transconductance Amplifier (OTA) With Power Capability" by L, Kaplan and H. Wittlinger, IEEE Transactions on Broadcast and Television Receivers, Vol. BTR-18, No.3, August, 1972.

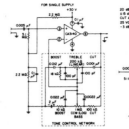

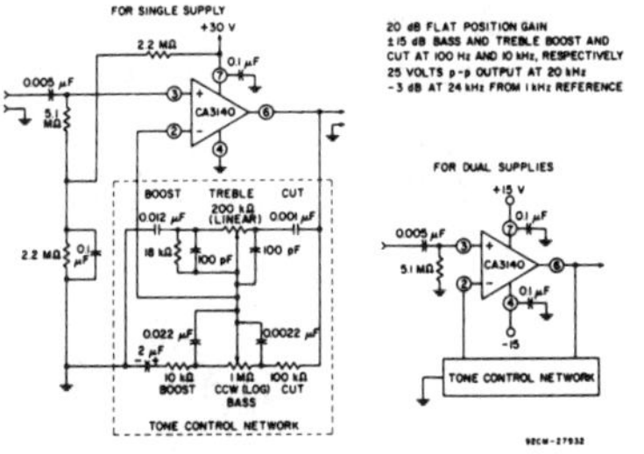

Fig. 30 – Tone control circuit using CA3130 series (20-dB midband gain).

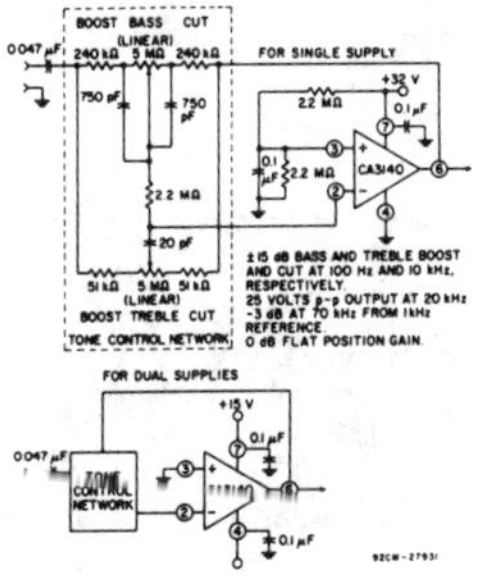

Fig. 31 – Baxandall tone control circuit using CA3140 series.

WIEN BRIDGE OSCILLATOR

Another application of the CA3140 that makes excellent use of its high input-impedance, high-slew-rate, and high-voltage qualities is the Wien Bridge sine-wave oscillator. A basic Wien Bridge oscillator is shown in Fig. 32. When $R_1 = R_2 = R$ and $C_1 = C_2 = C$, the frequency equation reduces to the familiar $f = 1/2\pi RC$ and the gain required for oscillation, A_{OSC}, is equal to 3. Note that if C_2 is increased by a factor of four and R_2 is reduced by a factor of four, the gain required for oscillation becomes 1.5, thus permitting a potentially higher operating frequency closer to the gain-bandwidth product of the CA3140.

Oscillator stabilization takes on many forms. It must be precisely set, otherwise the amplitude will either diminish or reach some form of limiting with high levels of distortion. The element, R_s, is commonly replaced with some variable resistance element. Thus, through some control means, the value of R_s is adjusted to maintain constant oscillator output. A FET channel resistance, a thermistor, a lamp bulb, or other device whose resistance is made to increase as the output amplitude is increased are a few of the elements often utilized.

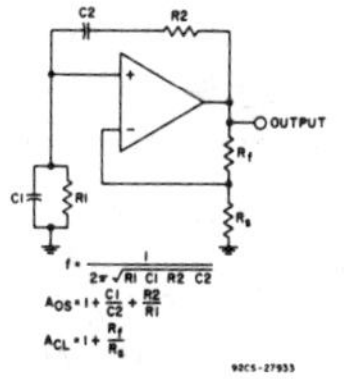

Fig. 32 – Basic Wien bridge oscillator circuit using an operational amplifier.

Fig. 33 shows another means of stabilizing the oscillator with a zener diode shunting the feedback resistor (R_f of Fig. 32). As the output signal amplitude increases, the zener diode impedance decreases resulting in more feedback with consequent reduction in gain; thus stabilizing the amplitude of the output signal. Furthermore, this combination of a monolithic zener diode and bridge rectifier circuit tends to provide a zero temperature coefficient for this regulating system. Because this bridge rectifier system has no time constant, i.e., thermal time constant for the lamp bulb, and RC time constant for filters often used in detector networks, there is no lower frequency limit. For example, with 1-μF polycarbonate capacitors and 22 MΩ for the frequency determining network, the operating frequency is 0.007 Hz.

As the frequency is increased, the output amplitude must be reduced to prevent the output signal from becoming slew-rate limited. An output frequency of 180 kHz will reach a slew rate of approximately 9 volts/μs when its amplitude is 16 volts peak-to-peak.

A-1. CA3140, CMOS (RCA)

CA3140, CA3140A, CA3140B Types

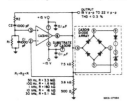

Fig. 33 — Wien bridge oscillator circuit using CA3140 series.

SIMPLE SAMPLE-AND-HOLD SYSTEM

Fig. 34 shows a very simple sample-and-hold system using the CA3140 as the readout amplifier for the storage capacitor. The CA3080A serves as both input buffer amplifier and low feed-through transmission switch.* System offset nulling is accomplished with the CA3140 via its offset nulling terminals. A typical simulated load of 2 kΩ and 30 pF is shown in the schematic.

In this circuit, the storage compensation capacitance (C_1) is only 200 pF. Larger value capacitors provide longer "hold" periods but with slower slew rates. The slew rate

$$\frac{dv}{dt} = \frac{i}{c} = 0.5 \text{ mA}/200 \text{ pF} = 2.5 \text{ V}/\mu s.$$

* ICAN-6668 "Applications of the CA3080 and CA3080A High-Performance Operational Transconductance Amplifiers".

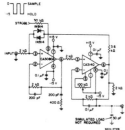

Fig. 34 — Sample- and hold circuit.

Pulse "droop" during the hold interval is 170 pA/200 pF which is = 0.85 μV/μs; (i.e., 170 pA/200 pF). In this case, 170 pA represents the typical leakage current of the CA3080A when strobed off. If C_1 were increased to 2000 pF, the "hold-droop" rate will decrease to 0.085 μV/μs, but the slew rate would decrease to 0.25 V/μs. The parallel diode network connected between terminal 3 of the CA3080A and terminal 6 of the CA3140 prevents large input-signal feed-through across the input terminals of the CA3080A to the 200 pF storage capacitor when the CA3080A is strobed off. Fig. 35 shows dynamic characteristic waveforms of this sample-and-hold system.

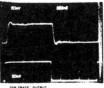

TOP TRACE: OUTPUT
(50 mV/DIV AND 200 ns/DIV.)
BOTTOM TRACE: INPUT
(50 mV/DIV AND 200 ns/DIV.)

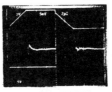

LARGE-SIGNAL RESPONSE AND SETTLING TIME
TOP TRACE: OUTPUT SIGNAL
(5 V/DIV AND 2μs/DIV.)
BOTTOM TRACE: INPUT SIGNAL
(5 V/DIV AND 2μs/DIV.)
CENTER TRACE: DIFFERENCE OF INPUT AND OUTPUT SIGNALS THROUGH TEKTRONIX AMPLIFIER 7A13
(5 mV/DIV AND 2μs/DIV.)

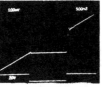

SAMPLING RESPONSE
TOP TRACE: SYSTEM OUTPUT
(100 mV/DIV AND 500 ns/DIV.)
BOTTOM TRACE: SAMPLING SIGNAL
(20 V/DIV AND 500 ns/DIV.)

Fig. 35 — Sample- and hold system dynamic characteristics waveforms.

CURRENT AMPLIFIER

The low input-terminal current needed to drive the CA3140 makes it ideal for use in current-amplifier applications such as the one shown in Fig. 36.● In this circuit, low current is supplied at the input potential as the power supply to load resistor R_L. This load current is increased by the multiplication factor R2/R1, when the load current is monitored by the power supply meter M. Thus, if the load current is 100 nA, with values shown, the load current presented to the supply will be 100 μA; a much easier current to measure in many systems.

Note that the input and output voltages are transferred at the same potential and only the output current is multiplied by the scale factor.

The dotted components show a method of decoupling the circuit from the effects of high output-load capacitance and the potential oscillation in this situation. Essentially, the necessary high-frequency feedback is provided by the capacitor with the dotted series resistor providing load decoupling.

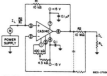

Fig. 36 — Basic current amplifier for low-current measurement systems.

Fig. 37 shows a single-supply, absolute-value, ideal full-wave rectifier with associated waveforms. During positive excursions, the input signal is fed through the feedback network directly to the output. Simultaneously, the positive excursion of the input signal also drives the output terminal (No.6) of the inverting amplifier in a negative-going excursion such that the 1N914 diode effectively disconnects the amplifier from the signal path. During a negative-going excursion of the input signal, the CA3140 functions as a normal inverting amplifier with a gain equal to —R2/R1. When the equality of the two equations shown in Fig. 37 is satisfied, the full-wave output is symmetrical.

● "Operational Amplifiers Design and Applications", J. G. Graeme, McGraw-Hill Book Company, page 308 — "Negative Immittance Converter Circuits".

A-1. CA3140, CMOS (RCA)

CA3140, CA3140A, CA3140B Types

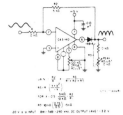
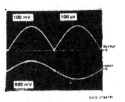

Fig. 37 – Single-supply, absolute-value, ideal full-wave rectifier with associated waveforms.

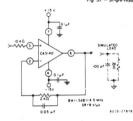
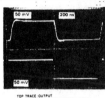
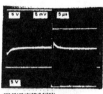

Fig. 38 – Split-supply voltage-follower test circuit and associated waveforms.

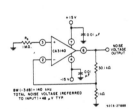
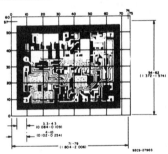

Fig. 39 – Test circuit amplifier (30-dB gain) used for wideband noise measurement.

CA3140H Chip

The photographs and dimensions represent a chip when it is part of the wafer. When the wafer is cut into chips, the cleavage angles are 57° instead of 90° with respect to the face of the chip. Therefore, the isolated chip is actually 7 mils (0.17 mm) larger in both dimensions.

Dimensions in parentheses are in millimeters and are derived from the basic inch dimensions as indicated. Grid graduations are in mils (10^{-3} inch).

A-1. CA3140, CMOS (RCA)

CA3141E

High-Voltage Diode Array
For Commercial, Industrial, and Military Applications

Features:
- Matched monolithic construction — V_F for each diode pair matched to within 0.55 mV (typ.) at I_F = 1 mA
- Low diode capacitance — 0.3 pF (typ.) at V_R = 2 V
- High diode-to-substrate breakdown voltage — 30 V (min.)
- Low reverse (leakage) current — 100 nA (max.)

Applications:
- Balanced modulators or demodulators
- Analog switches
- High-voltage diode gates
- Current ratio detectors

Fig. 1 — Terminal assignment.

The RCA-CA3141E High-Voltage Diode Array consists of ten general-purpose high-reverse-breakdown diodes. Six diodes are internally connected to form three common-cathode diode pairs, and the remaining four diodes are internally connected to form two common-anode diode pairs. Integrated circuit construction assures excellent static and dynamic matching of the diodes, making the CA3141E extremely useful for a wide variety of applications in communications and switching systems.

The CA3141E is supplied in the 16-lead dual-in-line plastic package (E suffix), and in chip form (H suffix).

MAXIMUM RATINGS, Absolute Maximum Values

PEAK INVERSE VOLTAGE (PIV)	30 V
PEAK DIODE-TO-SUBSTRATE VOLTAGE	30 V
PEAK FORWARD SURGE CURRENT [I_F (SURGE)]	100 mA
DC FORWARD CURRENT (I_F)	25 mA
DISSIPATION:	
Any one diode unit	50 mW
Total Package:	
Up to 55°C	650 mW
For T_A > 55°C	Derate linearly at 6.67 mW/°C
AMBIENT TEMPERATURE RANGE:	
Operating	−55 to +125°C
Storage	−65 to +150°C
LEAD TEMPERATURE (During Soldering):	
At distance 1/16 ± 1/32 inch (1.59 ± 0.79 mm) from case for 10s max.	+265°C

ELECTRICAL CHARACTERISTICS at T_A = 25°C

CHARACTERISTIC	TEST CONDITIONS	LIMITS Min.	LIMITS Typ.	LIMITS Max.	UNIT
DC Forward Voltage Drop, V_F	I_F (Anode) 100 μA	–	0.7	0.9	V
	1 mA	–	0.78	1	
	10 mA	–	0.93	1.2	
DC Reverse Breakdown Voltage, $V_{(BR)R}$	I_F = −10 μA	30	50	–	V
DC Breakdown Voltage Between Any Diode and Substrate, $V_{(BR)DI}$	I_{DI} = 10 μA	30	50	–	V
DC Reverse (Leakage) Current, I_R	V_F = −20 V	–	–	100	nA
DC Reverse (Leakage) Current Between Any Diode and Substrate, I_{DI}	V_{DI} = 20 V	–	–	100	nA
Magnitude of Diode Offset Voltage Between Diode Pairs	V_{DI} = 20 V, I_{FA} = 1 mA	–	0.55	–	mV
Temperature Coefficient of Forward Voltage Drop, $\Delta V_F/\Delta T$	I_F = 1 mA	–	−1.5	–	mV/°C
Reverse Recovery Time, t_{rr}	I_F = 2 mA, I_R = 2 mA	–	50	–	ns
Diode Capacitance, C_D			See Fig. 5		pF
Diode Anode-to-Substrate Capacitance, C_{DAI}			See Fig. 6		pF
Diode Cathode-to-Substrate Capacitance, C_{DCI}			See Fig. 7		pF
Magnitude of Cathode-to-Anode Current Ratio, $|I_{FC}/I_{FA}|$	I_{FA} = 1 mA, V_{DS} = 10 V	0.9	0.96	–	

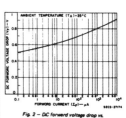

Fig. 2 — DC forward voltage drop vs. forward current.

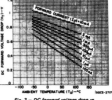

Fig. 3 — DC forward voltage drop vs. ambient temperature.

A-1. CA3140, CMOS (RCA)

uA747M, uA747C
DUAL GENERAL-PURPOSE OPERATIONAL AMPLIFIERS

D971, FEBRUARY 1971 – REVISED NOVEMBER 1988

- No Frequency Compensation Required
- Low Power Consumption
- Short-Circuit Protection
- Offset-Voltage Null Capability
- Wide Common-Mode and Differential Voltage Ranges
- No Latch-Up
- Designed to be Interchangeable with Fairchild µA747M and µA747C

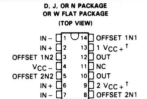

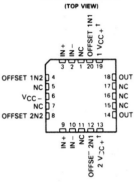

description

The uA747 is a dual general-purpose operational amplifier featuring offset-voltage null capability. Each half is electrically similar to uA741.

The high common-mode input voltage range and the absence of latch-up make this amplifier ideal for voltage-follower applications. The device is short-circuit protected and the internal frequency compensation ensures stability without external components. A low-value potentiometer may be connected between the offset null inputs to null out the offset voltage as shown in Figure 2.

The uA747M is characterized for operation over the full military temperature range of −55°C to 125°C; the uA747C is characterized for operation from 0°C to 70°C.

NC – No internal connection

†The two positive supply terminals (1 V_{CC+} and 2 V_{CC+}) are connected together internally.

symbol (each amplifier)

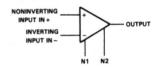

PRODUCTION DATA documents contain information current as of publication date. Products conform to specifications per the terms of Texas Instruments standard warranty. Production processing does not necessarily include testing of all parameters.

Copyright © 1983, Texas Instruments Incorporated

TEXAS INSTRUMENTS
POST OFFICE BOX 655012 • DALLAS, TEXAS 75265

A-2. µA747, dual general purpose (Texas Instruments)

uA747M, uA747C
DUAL GENERAL-PURPOSE OPERATIONAL AMPLIFIERS

schematic (each amplifier)

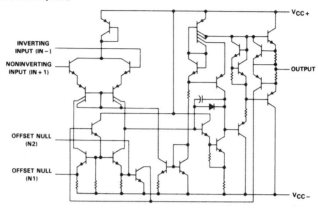

absolute maximum ratings over operating free-air temperature range (unless otherwise noted)

		uA747M	uA747C	UNIT
Supply voltage, V_{CC+} (see Note 1)		22	18	V
Supply voltage, V_{CC-} (see Note 1)		−22	−18	V
Differential input voltage (see Note 2)		±30	±30	V
Input voltage any input (see Notes 1 and 3)		±15	±15	V
Voltage between any offset null terminal (N1/N2) and V_{CC-}		±0.5	±0.5	V
Duration of output short-circuit (see Note 4)		unlimited	unlimited	
Continuous total dissipation		See Dissipation Rating Table		
Operating free-air temperature range		−55 to 125	0 to 70	°C
Storage temperature range		−65 to 150	−65 to 150	°C
Case temperature for 60 seconds	FK package	260		°C
Lead temperature 1,6 mm (1/16 inch) from case for 60 seconds	J or W package	300	300	°C
Lead temperature 1,6 mm (1/16 inch) from case for 10 seconds	D or N package		260	°C

NOTES: 1. All voltage values, unless otherwise noted, are with respect to the midpoint between V_{CC+} and V_{CC-}.
2. Differential voltages are at the noninverting input terminal with respect to the inverting input terminal.
3. The magnitude of the input voltage must never exceed the magnitude of the supply voltage or 15 volts, whichever is less.
4. The output may be shorted to ground or either power supply. For the uA747M only, the unlimited duration of the short-circuit applies at (or below) 125 °C case temperature or 75 °C free-air temperature.

DISSIPATION RATING TABLE

PACKAGE	$T_A \leq 25°C$ POWER RATING	DERATING FACTOR	DERATE ABOVE T_A	$T_A = 70°C$ POWER RATING	$T_A = 125°C$ POWER RATING
D	800 mW	7.6 mW/°C	45 °C	608 mW	−
FK	800 mW	11.0 mW/°C	77 °C	800 mW	275 mW
J (uA747M)	800 mW	11.0 mW/°C	77 °C	800 mW	275 mW
J (uA747C)	800 mW	8.2 mW/°C	52 °C	656 mW	−
N	800 mW	9.2 mW/°C	63 °C	736 mW	−
W	800 mW	8.0 mW/°C	50 °C	640 mW	−

TEXAS INSTRUMENTS
POST OFFICE BOX 655012 • DALLAS, TEXAS 75265

A-2. µA747, dual general purpose (Texas Instruments)

ua747M, ua747C
DUAL GENERAL-PURPOSE OPERATIONAL AMPLIFIERS

electrical characteristics at specified free-air temperature, $V_{CC+} = 15$ V, $V_{CC-} = -15$ V

PARAMETER		TEST CONDITIONS†		uA747M			uA747C			UNIT
				MIN	TYP	MAX	MIN	TYP	MAX	
V_{IO}	Input offset voltage	$V_O = 0$ V	25°C		1	5		1	6	mV
			Full range			6			7.5	
$\Delta V_{IO(adj)}$	Offset voltage adjust range		25°C		±15			±15		mV
I_{IO}	Input offset current		25°C		20	200		20	200	nA
			Full range			500			300	
I_{IB}	Input bias current		25°C		80	500		80	500	nA
			Full range			1500			800	
V_{ICR}	Common-mode input voltage range		25°C	±12	±13		±12	±13		V
			Full range	±12			±12			
V_{OPP}	Maximum peak-to-peak output voltage swing	$R_L = 10$ kΩ	25°C		24	28		24	28	V
		$R_L \geq 10$ kΩ	Full range	24			24			
		$R_L = 2$ kΩ	25°C	20	26		20	26		
		$R_L \geq 2$ kΩ	Full range	20			20			
A_{VD}	Large-signal differential voltage amplification	$R_L \geq 2$ kΩ, $V_O = ±10$ V	25°C	50	200		25	200		V/mV
			Full range	25			15			
r_i	Input resistance		25°C	0.3	2		0.3	2		MΩ
r_o	Output resistance	See Note 6	25°C		75			75		Ω
C_i	Input capacitance		25°C		1.4			1.4		pF
CMRR	Common-mode rejection ratio	$V_{IC} = V_{ICR}$	25°C	70	90		70	90		dB
			Full range	70			70			
k_{SVS}	Supply voltage sensitivity ($\Delta V_{IO}/\Delta V_{CC}$)	$V_{CC} = ±9$ V to ±15 V	25°C		30	150		30	150	μV/V
			Full range			150			150	
I_{OS}	Short-circuit output current		25°C		±25	±40		±20	±40	mA
I_{CC}	Supply current (each amplifier)	No load	25°C		1.7	2.8		1.7	2.8	mA
			Full range			3.3			3.3	
P_D	Power dissipation (each amplifier)	No load, $V_O = 0$ V	25°C		50	85		50	85	mW
			Full range			100			100	
V_{o1}/V_{o2}	Channel separation		25°C		120	0		120		dB

†All characteristics are measured under open-loop conditions with zero common-mode input voltage unless otherwise specified. Full range for uA747M is −55°C to 125°C and for uA747C is 0°C to 70°C.

NOTE 6: This typical value applies only at frequencies above a few hundred hertz because of the effects of drift and thermal feedback.

operating characteristics, $V_{CC+} = 15$ V, $V_{CC-} = -15$ V, $T_A = 25$°C

	PARAMETER	TEST CONDITIONS	uA747M			uA747C			UNIT
			MIN	TYP	MAX	MIN	TYP	MAX	
t_r	Rise time	$V_I = 20$ mV, $R_L = 2$ kΩ, $C_L = 100$ pF, See Figure 1		0.3			0.3		μs
	Overshoot factor			5%			5%		
SR	Slew rate at unity gain	$V_I = 10$ V, $R_L = 2$ kΩ, $C_L = 100$ pF, See Figure 1		0.5			0.5		V/μs

TEXAS INSTRUMENTS
POST OFFICE BOX 655012 • DALLAS, TEXAS 75265

A-2. μA747, dual general purpose (Texas Instruments)

uA747M, uA747C
DUAL GENERAL-PURPOSE OPERATIONAL AMPLIFIERS

PARAMETER MEASUREMENT INFORMATION

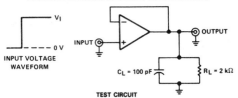

FIGURE 1. RISE TIME, OVERSHOOT, AND SLEW RATE

TYPICAL APPLICATION DATA

FIGURE 2. INPUT OFFSET VOLTAGE NULL CIRCUIT

TYPICAL CHARACTERISTICS

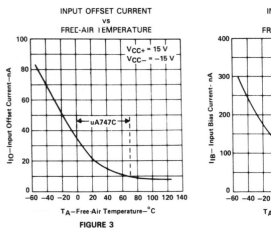

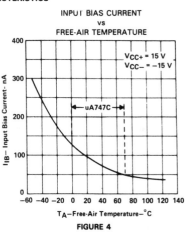

FIGURE 3

FIGURE 4

POST OFFICE BOX 655012 • DALLAS, TEXAS 75265

A-2. µA747, dual general purpose (Texas Instruments)

uA747M, uA747C
DUAL GENERAL-PURPOSE OPERATIONAL AMPLIFIERS

TYPICAL CHARACTERISTICS

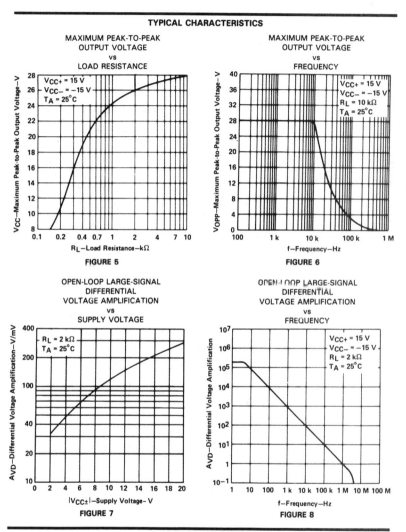

FIGURE 5 — MAXIMUM PEAK-TO-PEAK OUTPUT VOLTAGE vs LOAD RESISTANCE

FIGURE 6 — MAXIMUM PEAK-TO-PEAK OUTPUT VOLTAGE vs FREQUENCY

FIGURE 7 — OPEN-LOOP LARGE-SIGNAL DIFFERENTIAL VOLTAGE AMPLIFICATION vs SUPPLY VOLTAGE

FIGURE 8 — OPEN-LOOP LARGE-SIGNAL DIFFERENTIAL VOLTAGE AMPLIFICATION vs FREQUENCY

TEXAS INSTRUMENTS
POST OFFICE BOX 655012 • DALLAS, TEXAS 75265

A-2. µA747, dual general purpose (Texas Instruments)

uA747M, uA747C
DUAL GENERAL-PURPOSE OPERATIONAL AMPLIFIERS

TYPICAL CHARACTERISTICS

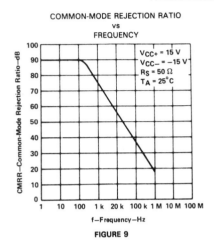

FIGURE 9

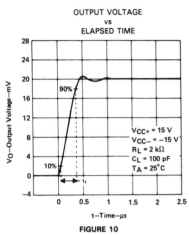

FIGURE 10

FIGURE 11

A-2. µA747, dual general purpose *(Texas Instruments)*

LF351
WIDE-BANDWIDTH JFET-INPUT OPERATIONAL AMPLIFIER

D2997, MARCH 1987

- Low Input Bias Current
  Typically 50 pA
- Low Input Noise Voltage
  Typically 18 nV/$\sqrt{Hz}$
- Low Input Noise Current
  Typically 0.01 pA/$\sqrt{Hz}$
- Low Supply Current . . . Typically 1.8 mA
- High Input Impedance
  Typically 10^{12} Ω
- Low Total Harmonic Distortion
- Internally Trimmed Offset Voltage
  Typically 10 mV
- High Slew Rate . . . Typically 13 V/µs
- Wide Gain Bandwidth . . . Typically 3 MHz
- Pin Compatible with Standard 741

P, D, OR JG PACKAGE
(TOP VIEW)

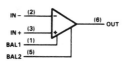

NC — No internal connection

description

This device is a low-cost, high-speed, JFET-input operational amplifier with an internally trimmed input offset voltage. It requires low supply current yet maintains a large gain-bandwidth product and a fast slew rate. In addition, the matched high-voltage JFET input provides very low input bias and offset currents. It uses the same offset voltage adjustment circuits as the 741.

The LF351 can be used in applications such as high-speed integrators, digital-to-analog converters, sample-and-hold circuits, and many other circuits.

The LF351 is characterized for operation from 0°C to 70°C.

symbol (each amplifier)

AVAILABLE OPTIONS

DEVICE	SYMBOLIZATION PACKAGE SUFFIX	OPERATING TEMPERATURE RANGE	V_{IO} MAX at 25°C
LF351	D,JG,P	–0°C to 70°C	10 mV

The D packages are available taped and reeled. Add the suffix R to the device type when ordering. (ie., LF351DR)

PRODUCTION DATA documents contain information current as of publication date. Products conform to specifications per the terms of Texas Instruments standard warranty. Production processing does not necessarily include testing of all parameters.

TEXAS INSTRUMENTS
POST OFFICE BOX 655012 • DALLAS, TEXAS 75265

Copyright © 1987, Texas Instruments Incorporated

A-3. LF351, JFET-input (Texas Instruments)

LF351
WIDE-BANDWIDTH JFET-INPUT OPERATIONAL AMPLIFIER

absolute maximum ratings over operating free-air temperature range (unless otherwise noted)

Supply voltage, V_{CC+}	18 V
Supply voltage, V_{CC-}	−18 V
Differential input voltage, V_{ID}	±30 V
Input voltage (see Note 1)	±15 V
Duration of output short circuit	Unlimited
Continuous total power dissipation	500 mW
Operating temperature range	0°C to 70°C
Storage temperature range	−65°C to 150°C
Lead temperature 1,6 mm (1/16 inch) from case for 60 seconds, JG package	300°C
Lead temperature 1,6 mm (1/16 inch) from case for 10 seconds, D or P package	260°C

NOTE 1: Unless otherwise specified, the absolute maximum negative input voltage is equal to the negative power supply voltage.

electrical characteristics over operating free-air temperature range, V_{CC+} = 15 V, V_{CC-} = −15 V (unless otherwise specified)

PARAMETER		TEST CONDITIONS		MIN	TYP	MAX	UNIT
V_{IO}	Input offset voltage	$V_{IC} = 0$, $R_S = 10$ kΩ	$T_A = 25°C$		5	10	mV
			Full range			13	
α_{VIO}	Average temperature coefficient of input offset voltage	$V_{IC} = 0$, $R_S = 10$ kΩ			10		μV/°C
I_{IO}	Input offset current†	$V_{IC} = 0$	$T_J = 25°C$		25	100	pA
			$T_J = 70°C$			4	nA
I_{IB}	Input bias current†	$V_{IC} = 0$	$T_J = 25°C$		50	200	pA
			$T_J = 70°C$			8	nA
V_{ICR}	Common-mode input voltage range			±11	−12 to 15		V
V_{OM}	Maximum peak output voltage swing	$R_L = 10$ kΩ		±12	±13.5		V
A_{VD}	Large-signal differential voltage	$V_O = ±10$ V, $R_L = 2$ kΩ	$T_A = 25°C$	25	200		V/mV
			Full range	15	200		
r_i	Input resistance	$T_J = 25°C$			10^{12}		Ω
CMRR	Common-mode rejection ratio	$R_S ≤ 10$ kΩ		70	100		dB
k_{SVR}	Supply voltage rejection ratio	See Note 2		70	100		dB
I_{CC}	Supply current				1.8	3.4	mA

operating characteristics, V_{CC+} = 15 V, V_{CC-} = −15 V, T_A = 25°C

PARAMETER		TEST CONDITIONS	MIN	TYP	MAX	UNIT
SR	Slew rate		8	13		V/μs
B_1	Unity-gain bandwidth			3		MHz
V_n	Equivalent input noise voltage	f = 1 kHz, $R_S = 100$ Ω		18		nV/√Hz
I_n	Equivalent input noise current	f = 1 kHz		0.01		pA/√Hz

† Input bias currents of a FET-input operational amplifier are normal junction reverse currents, which are temperature sensitive. Pulse techniques must be used that will maintain the junction temperatures as close to the ambient temperature as possible.

NOTE 2: Supply voltage rejection ratio is measured for both supply magnitudes increasing or decreasing simultaneously.

A-3. LF351, JFET-input (Texas Instruments)

LM101A, LM201A, LM301A
HIGH-PERFORMANCE OPERATIONAL AMPLIFIERS

D961, OCTOBER 1979 – REVISED JUNE 1988

- Low Input Currents
- Low Input Offset Parameters
- Frequency and Transient Response Characteristics Adjustable
- Short-Circuit Protection
- Offset-Voltage Null Capability
- No Latch-Up
- Wide Common-Mode and Differential Voltage Ranges
- Same Pin Assignments as uA709
- Designed to be Interchangeable with National Semiconductor LM101A and LM301A

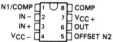

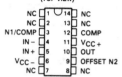

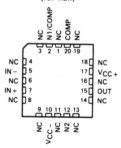

NC – No internal connection

description

The LM101A, LM201A, and LM301A are high-performance operational amplifiers featuring very low input bias current and input offset voltage and current to improve the accuracy of high-impedance circuits using these devices. The high common-mode input voltage range and the absence of latch-up make these amplifiers ideal for voltage-follower applications. The devices are protected to withstand short circuits at the output. The external compensation of these amplifiers allows the changing of the frequency response (when the closed-loop gain is greater than unity) for applications requiring wider bandwidth or higher slew rate. A potentiometer may be connected between the offset-null inputs (N1 and N2), as shown in Figure 7, to null out the offset voltage.

The LM101A is characterized for operation over the full military temperature range of −55°C to 125°C, the LM201A is characterized for operation from −25°C to 85°C, and the LM301A is characterized for operation from 0°C to 70°C.

symbol

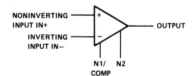

POST OFFICE BOX 655012 • DALLAS, TEXAS 75265

Copyright © 1983 Texas Instruments Incorporated

A-4. *LM101A/201A/301A, high-performance* (Texas Instruments)

261

LM101A, LM201A, LM301A
HIGH-PERFORMANCE OPERATIONAL AMPLIFIERS

AVAILABLE OPTIONS

T_A	V_{IO} MAX at 25°C	PACKAGE					
		SMALL OUTLINE (D)	CHIP CARRIER (FK)	CERAMIC DIP (JG)	PLASTIC DIP (P)	FLAT PACK (U)	FLAT PACK (W)
0°C to 70°C	7.5 mV	LM301AD	–	LM301AJG	LM301AP	–	–
–25°C to 85°C	2 mV	LM201AD	–	LM201AJG	LM201AP	–	–
–55°C to 125°C	2 mV	–	LM101AFK	LM101AJG	–	LM101AU	LM101AW

The D package is available taped and reeled. Add the suffix R to the device type. (i.e., LM301ADR)

absolute maximum ratings over operating free-air temperature range (unless otherwise noted)

	LM101A	LM201A	LM301A	UNIT
Supply voltage V_{CC+} (see Note 1)	22	22	18	V
Supply voltage V_{CC-} (see Note 1)	–22	–22	–18	V
Differential input voltage (see Note 2)	±30	±30	±30	V
Input voltage (either input, see Notes 1 and 3)	±15	±15	±15	V
Voltage between either offset null terminal (N1/N2) and V_{CC-}	–0.5 to 2	–0.5 to 2	–0.5 to 2	V
Duration of output short-circuit (see Note 4)	unlimited	unlimited	unlimited	
Continuous total power dissipation	See Dissipation Rating Table			
Operating free-air temperature range	–55 to 125	–25 to 85	0 to 70	°C
Storage temperature range	–65 to 150	–65 to 150	–65 to 150	°C
Case temperature for 60 seconds: FK package	260			°C
Lead temperature 1,6 mm (1/16 inch) from case for 60 seconds — JG, U, or W package	300	300	300	°C
Lead temperature 1,6 mm (1/16 inch) from case for 10 seconds — D or P package		260	260	°C

NOTES: 1. All voltage values, unless otherwise noted, are with respect to the midpoint between V_{CC+} and V_{CC-}.
2. Differential voltages are at the noninverting input terminal with respect to the inverting input terminal.
3. The magnitude of the input voltage must never exceed the magnitude of the supply voltage or 15 V, whichever is less.
4. The output may be shorted to ground or either power supply. For the LM101A only, the unlimited duration of the short-circuit applies at (or below) 125°C case temperature or 75°C free-air temperature. For the LM201A only, the unlimited duration of the short-circuit applies at (or below) 85°C case temperature or 75°C free-air temperature.

DISSIPATION RATING TABLE

PACKAGE	$T_A \leq 25°C$ POWER RATING	DERATING FACTOR	DERATE ABOVE T_A	$T_A = 70°C$ POWER RATING	$T_A = 85°C$ POWER RATING	$T_A = 125°C$ POWER RATING
D	500 mW	5.8 mW/°C	64°C	464 mW	377 mW	N/A
FK	500 mW	11.0 mW/°C	105°C	500 mW	500 mW	275 mW
JG (LM101A)	500 mW	8.4 mW/°C	90°C	500 mW	500 mW	210 mW
JG (LM201A, LM301A)	500 mW	6.6 mW/°C	74°C	500 mW	429 mW	N/A
P	500 mW	N/A	N/A	500 mW	500 mW	N/A
U	500 mW	5.4 mW/°C	57°C	432 mW	351 mW	135 mW
W	500 mW	8.0 mW/°C	87°C	500 mW	500 mW	200 mW

A-4. LM101A/201A/301A, high-performance (Texas Instruments)

LM101A, LM201A, LM301A
HIGH-PERFORMANCE OPERATIONAL AMPLIFIERS

electrical characteristics at specified free-air temperature, C_C = 30 pF (see Note 5)

PARAMETER		TEST CONDITIONS†		LM101A, LM201A			LM301A			UNIT
				MIN	TYP	MAX	MIN	TYP	MAX	
V_{IO}	Input offset voltage	V_O = 0 V	25°C		0.6	2		2	7.5	mV
			Full range			3			10	
α_{VIO}	Average temperature coefficient of input offset voltage	V_O = 0 V	Full range		3	15		6	30	µV/°C
I_{IO}	Input offset current		25°C		1.5	10		3	50	nA
			Full range			20			70	
α_{IIO}	Average temperature coefficient of input offset current	T_A = −55°C to 25°C			0.02	0.2				nA/°C
		T_A = 25°C to MAX			0.01	0.1				
		T_A = 0°C to 25°C						0.02	0.6	
		T_A = 25°C to 70°C						0.01	0.3	
I_{IB}	Input bias current		25°C		30	75		70	250	nA
			Full range			100			300	
V_{ICR}	Common-mode input voltage range	See Note 6	Full range	±15			±12			V
V_{OPP}	Maximum peak-to-peak output voltage swing	$V_{CC\pm}$ = ±15 V, R_L = 10 kΩ	25°C	24	28		24	28		V
			Full range	24			24			
		$V_{CC\pm}$ = ±15 V, R_L = 2 kΩ	25°C	20	26		20	26		
			Full range	20			20			
A_{VD}	Large-signal differential voltage amplification	$V_{CC\pm}$ = ±15 V, V_O = ±10 V, R_L ≥ 2 kΩ	25°C	50	200		25	200		V/mV
			Full range	25			15			
r_i	Input resistance		25°C	1.5	4		0.5	2		MΩ
CMRR	Common-mode rejection ratio	V_{IC} = V_{ICR} min	25°C	80	98		70	90		dB
			Full range	80			70			
k_{SVR}	Supply voltage rejection ratio (Δ V_{CC}/Δ V_{IO})		25°C	80	98		70	96		dB
			Full range	80			70			
I_{CC}	Supply current	No load, V_O = 0, See Note 6	25°C		1.8	3		1.8	3	mA
			MAX		1.2	2.5				

† All characteristics are measured under open-loop conditions with zero common-mode input voltage unless otherwise specified. Full range for LM101A is −55°C to 125°C, for LM201A is −25°C to 85°C, and for LM301A is 0°C to 70°C.

NOTES: 5. Unless otherwise noted, $V_{CC\pm}$ = ±5 to ±20 V for LM101A and LM201A, and $V_{CC\pm}$ = ±5 V to ±15 V for LM301A. All typical values are at $V_{CC\pm}$ = ±15 V.
6. For LM101A and LM201A, $V_{CC\pm}$ = ±20 V. For LM301A, $V_{CC\pm}$ = ±15 V.

TEXAS INSTRUMENTS
POST OFFICE BOX 655012 • DALLAS, TEXAS 75265

A-4. LM101A/201A/301A, high-performance (Texas Instruments)

LM101A, LM201A, LM301A
HIGH-PERFORMANCE OPERATIONAL AMPLIFIERS

TYPICAL CHARACTERISTICS

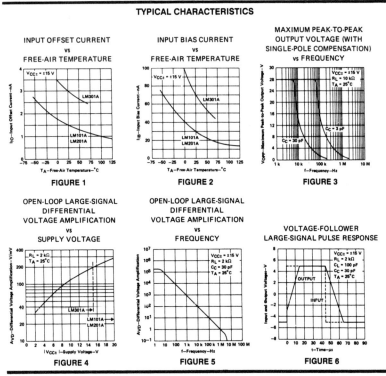

FIGURE 1. INPUT OFFSET CURRENT vs FREE-AIR TEMPERATURE

FIGURE 2. INPUT BIAS CURRENT vs FREE-AIR TEMPERATURE

FIGURE 3. MAXIMUM PEAK-TO-PEAK OUTPUT VOLTAGE (WITH SINGLE-POLE COMPENSATION) vs FREQUENCY

FIGURE 4. OPEN-LOOP LARGE-SIGNAL DIFFERENTIAL VOLTAGE AMPLIFICATION vs SUPPLY VOLTAGE

FIGURE 5. OPEN-LOOP LARGE-SIGNAL DIFFERENTIAL VOLTAGE AMPLIFICATION vs FREQUENCY

FIGURE 6. VOLTAGE-FOLLOWER LARGE-SIGNAL PULSE RESPONSE

TYPICAL APPLICATION DATA

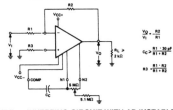

FIGURE 7. INVERTING CIRCUIT WITH ADJUSTABLE GAIN, SINGLE-POLE COMPENSATION, AND OFFSET ADJUSTMENT

POST OFFICE BOX 655012 • DALLAS, TEXAS 75265

A-4. LM101A/201A/301A, high-performance (Texas Instruments)

LM218, LM318
HIGH-PERFORMANCE OPERATIONAL AMPLIFIERS

D2219, JUNE 1976 – REVISED OCTOBER 1988

- Small-Signal Bandwidth . . . 15 MHz Typ
- Slew Rate . . . 50 V/µs Min
- Bias Current . . . 250 nA Max (LM218)
- Supply Voltage Range . . . ±5 V to ±20 V
- Internal Frequency Compensation
- Input and Output Overload Protection
- Same Pin Assignments as General-Purpose Operational Amplifiers

D, JG, OR P PACKAGE
(TOP VIEW)

```
BAL/COMP  [1    8]  COMP 2
IN-       [2    7]  VCC+
IN+       [3    6]  OUT
VCC-      [4    5]  BAL/COMP 3
```

description

The LM218 and LM318 are precision, high-speed operational amplifiers designed for applications requiring wide bandwidth and high slew rate. They feature a factor-of-ten increase in speed over general purpose devices without sacrificing dc performance.

These operational amplifiers have internal unity-gain frequency compensation. This considerably simplifies their application since no external components are necessary for operation. However, unlike most internally compensated amplifiers, external frequency compensation may be added for optimum performance. For inverting applications, feed-forward compensation will boost the slew rate to over 150 V/µs and almost double the bandwidth. Over compensation may be used with the amplifier for greater stability when maximum bandwidth is not needed. Further, a single capacitor may be added to reduce the settling time for 0.1% error band to under 1 µs.

The high speed and fast settling time of these operational amplifiers make them useful in A/D converters, oscillators, active filters, sample and hold circuits, and general purpose amplifiers.

The LM218 is characterized for operation from −25°C to 85°C, and the LM318 is characterized for operation from 0°C to 70°C.

symbol

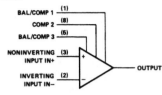

AVAILABLE OPTIONS

T_A	V_{IO} MAX at 25°C	PACKAGE		
		SMALL-OUTLINE (D)	CERAMIC DIP (JG)	PLASTIC DIP (P)
0°C to 70°C	10 mV	LM318D	LM318JG	LM318P
−25°C to 85°C	4 mV	LM218D	LM218JG	LM218P

The D packages are available taped and reeled. Add the suffix R to the device type (e.g., LM318DR).

PRODUCTION DATA documents contain information current as of publication date. Products conform to specifications per the terms of Texas Instruments standard warranty. Production processing does not necessarily include testing of all parameters.

Copyright © 1983, Texas Instruments Incorporated

TEXAS INSTRUMENTS
POST OFFICE BOX 655012 • DALLAS, TEXAS 75265

A-5. LM218/318, high-performance (Texas Instruments)

LM218, LM318
HIGH-PERFORMANCE OPERATIONAL AMPLIFIERS

schematic

A-5. *LM218/318, high-performance* (Texas Instruments)

LM218, LM318
HIGH-PERFORMANCE OPERATIONAL AMPLIFIERS

absolute maximum ratings over operating free-air temperature range (unless otherwise noted)

	LM218	LM318	UNIT
Supply voltage, V_{CC+} (see Note 1)	20	20	V
Supply voltage, V_{CC-} (see Note 1)	−20	−20	V
Input voltage (either input, see Notes 1 and 2)	±15	±15	V
Differential input current (see Note 3)	±10	±10	mA
Duration of output short-circuit (see Note 4)	unlimited	unlimited	
Continuous total power dissipation	See Dissipation Rating Table		
Operating free-air temperature range	−25 to 85	0 to 70	°C
Storage temperature range	−65 to 150	−65 to 150	°C
Lead temperature 1,6 mm (1/16 inch) from case for 60 seconds JG package	300	300	°C
Lead temperature 1,6 mm (1/16 inch) from case for 10 seconds D or P package	260	260	°C

NOTES: 1. All voltage values, unless otherwise noted, are with respect to the midpoint between V_{CC+} and V_{CC-}.
2. The magnitude of the input voltage must never exceed the magnitude of the supply voltage or 15 V, whichever is less.
3. The inputs are shunted with two opposite-facing base-emitter diodes for over voltage protection. Therefore, excessive current will flow if a differential input voltage in excess of approximately 1 V is applied between the inputs unless some limiting resistance is used.
4. The output may be shorted to ground or either power supply. For the LM218 only, the unlimited duration of the short-circuit applies at (or below) 85°C case temperature or 75°C free-air temperature.

DISSIPATION RATING TABLE

PACKAGE	$T_A \leq 25°C$ POWER RATING	DERATING FACTOR	DERATE ABOVE T_A	$T_A = 70°C$ POWER RATING	$T_A = 85°C$ POWER RATING
D	500 mW	5.8 mW/°C	64°C	464 mW	377 mW
JG	500 mW	6.6 mW/°C	74°C	500 mW	429 mW
P	500 mW	N/A	N/A	500 mW	500 mW

A-5. LM218/318, high-performance (Texas Instruments)

LM218, LM318
HIGH-PERFORMANCE OPERATIONAL AMPLIFIERS

electrical characteristics at specified free-air temperature (see Note 5)

	PARAMETER	TEST CONDITIONS†		LM218			LM318			UNIT
				MIN	TYP	MAX	MIN	TYP	MAX	
V_{IO}	Input offset voltage	$V_O = 0$	25°C		2	4		4	10	mV
			Full range			6			15	
I_{IO}	Input offset current	$V_O = 0$	25°C		6	50		30	200	nA
			Full range			100			300	
I_{IB}	Input bias current	$V_O = 0$	25°C		120	250		150	500	nA
			Full range			500			750	
V_{ICR}	Common-mode input voltage range	$V_{CC\pm} = \pm15$ V	Full range	±11.5			±11.5			V
V_{OM}	Maximum peak output voltage swing	$V_{CC\pm} = \pm15$ V, $R_L = 2$ kΩ	Full range	±12	±13		±12	±13		V
A_{VD}	Large-signal differential voltage amplification	$V_{CC\pm} = \pm15$ V, $V_O = \pm10$ V, $R_L \geq 2$ kΩ	25°C	50	200		25	200		V/mV
			Full range	25			20			
B_1	Unity-gain bandwidth	$V_{CC\pm} = \pm15$ V	25°C		15			15		MHz
r_i	Input resistance		25°C		3			0.5	3	MΩ
CMRR	Common-mode rejection ratio	$V_{IC} = V_{ICR}$ min	Full range	80	100		70	100		dB
k_{SVR}	Supply voltage rejection ratio ($\Delta V_{CC}/\Delta V_{IO}$)		Full range	70	80		65	80		dB
I_{CC}	Supply current	No load, $V_O = 0$	25°C		5	8		5	10	mA

† All characteristics are measured under open-loop conditions with zero common-mode input voltage unless otherwise specified. Full range for LM218 is −25°C to 85°C and for LM318 is 0°C to 70°C.
Note 5: Unless otherwise noted, $V_{CC} = \pm5$ V to ±20 V. All typical values are at $V_{CC\pm} = \pm15$ V, $T_A = 25$°C.

operating characteristics, $V_{CC+} = 15$ V, $V_{CC-} = 15$ V, $T_A = 25$°C

	PARAMETER	TEST CONDITIONS			MIN	TYP	MAX	UNIT
SR	Slew rate at unity gain	$\Delta V_I = 10$ V.	$C_L = 10$ pF.	See Figure 1		50	70	V/μs

PARAMETER MEASUREMENT INFORMATION

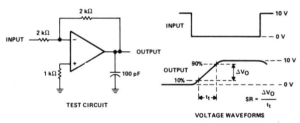

FIGURE 1. SLEW RATE

POST OFFICE BOX 655012 • DALLAS, TEXAS 75265

A-5. LM218/318, high-performance (Texas Instruments)

LM2900, LM3900
QUADRUPLE OPERATIONAL AMPLIFIERS

D2531, JULY 1979 – REVISED AUGUST 1988

- Wide Range of Supply Voltages, Single or Dual Supplies
- Wide Bandwidth
- Large Output Voltage Swing
- Output Short-Circuit Protection
- Internal Frequency Compensation
- Low Input Bias Current
- Designed to be Interchangeable with National Semiconductor LM2900 and LM3900, Respectively

J OR N DUAL-IN-LINE PACKAGE
(TOP VIEW)

```
#1 IN+  [1   14] VCC
#2 IN+  [2   13] #3 IN+
#2 IN-  [3   12] #4 IN+
#2 OUT  [4   11] #4 IN-
#1 OUT  [5   10] #4 OUT
#1 IN-  [6    9] #3 OUT
GND     [7    8] #3 IN-
```

description

These devices consist of four independent, high-gain frequency-compensated Norton operational amplifiers that were designed specifically to operate from a single supply over a wide range of voltages. Operation from split supplies is also possible. The low supply current drain is essentially independent of the magnitude of the supply voltage. These devices provide wide bandwidth and large output voltage swing.

The LM2900 is characterized for operation from −40°C to 85°C, and the LM3900 is characterized for operation from 0°C to 70°C.

AVAILABLE OPTIONS

T_A	PACKAGE	
	PLASTIC DIP (N)	CERAMIC DIP (J)
0°C to 70°C	LM3900N	LM3900J
−40°C to 85°C	LM2900N	LM2900J

symbol (each amplifier)

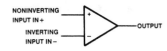

schematic (each amplifier)

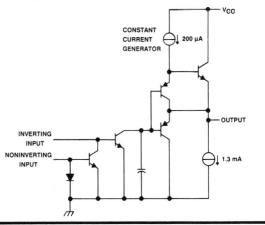

POST OFFICE BOX 655012 • DALLAS, TEXAS 75265

Copyright © 1979, Texas Instruments Incorporated

PRODUCTION DATA documents contain information current as of publication date. Products conform to specifications per the terms of Texas Instruments standard warranty. Production processing does not necessarily include testing of all parameters.

A-6. *LM2900/3900, quad Norton* (Texas Instruments)

LM2900, LM3900
QUADRUPLE OPERATIONAL AMPLIFIERS

absolute maximum ratings over operating free-air temperature range (unless otherwise noted)

		LM2900	LM3900	UNIT
Supply voltage, V_{CC} (see Note 1)		36	36	V
Input current		20	20	mA
Duration of output short circuit (one amplifier) to ground at (or below) 25°C free-air temperature (see Note 2)		unlimited	unlimited	
Continuous total dissipation		See Dissipation Rating Table		
Operating free-air temperature range		−40 to 85	0 to 70	°C
Storage temperature range		−65 to 150	−65 to 150	°C
Lead temperature 1,6 mm (1/16 inch) from case for 60 seconds	J Package	300	300	°C
Lead temperature 1,6 mm (1/16 inch) from case for 10 seconds	N Package	260	260	°C

NOTES: 1. All voltage values, except differential voltages, are with respect to the network ground terminal.
2. Short circuits from outputs to V_{CC} can cause excessive heating and eventual destruction.

DISSIPATION RATING TABLE

PACKAGE	$T_A \leq 25°C$ POWER RATING	DERATING FACTOR ABOVE $T_A = 25°C$	$T_A = 70°C$ POWER RATING	$T_A = 85°C$ POWER RATING
J	1025 mW	8.2 mW/°C	656 mW	533 mW
N	1150 mW	9.2 mW/°C	736 mW	598 mW

recommended operating conditions

	LM2900		LM3900		UNIT
	MIN	MAX	MIN	MAX	
Input current (see Note 3)		−1		−1	mA
Operating free-air temperature, T_A	−40	85	0	70	°C

NOTE 3: Clamp transistors are included that prevent the input voltages from swinging below ground more than approximately −0.3 V. The negative input currents that may result from large signal overdrive with capacitive input coupling must be limited externally to values of approximately −1 mA. Negative input currents in excess of −4 mA will cause the output voltage to drop to a low voltage. These values apply for any one of the input terminals. If more than one of the input terminals are simultaneously driven negative, maximum currents are reduced. Common-mode current biasing can be used to prevent negative input voltages.

TEXAS INSTRUMENTS
POST OFFICE BOX 655012 • DALLAS, TEXAS 75265

A-6. LM2900/3900, quad Norton (Texas Instruments)

LM2900, LM3900
QUADRUPLE OPERATIONAL AMPLIFIERS

electrical characteristics, V_{CC} = 15 V, T_A = 25°C (unless otherwise noted)

PARAMETER		TEST CONDITIONS†		LM2900 MIN	LM2900 TYP	LM2900 MAX	LM3900 MIN	LM3900 TYP	LM3900 MAX	UNIT
I_{IB}	Input bias current (inverting input)	$I_{I+} = 0$	$T_A = 25°C$		30	200		30	200	nA
			T_A = Full range			300			300	
I_{I-} / I_{I+}	Mirror gain	I_{I+} = 20 μA to 200 μA, T_A = Full range, See Note 4		0.9		1.1	0.9		1.1	μA/μA
	Change in mirror gain				2	5		2	15	%
	Mirror current	$V_{I+} = V_{I-}$, T_A = Full range, See Note 4			10	500		10	500	μA
A_{VD}	Large-signal differential voltage amplification	V_O = 10 V, R_L = 10 kΩ, f = 100 Hz		1.2	2.8		1.2	2.8		V/mV
r_i	Input resistance (inverting input)				1			1		MΩ
r_o	Output resistance				8			8		kΩ
B_1	Unity-gain bandwidth (inverting input)				2.5			2.5		MHz
k_{SVR}	Supply voltage rejection ratio ($\Delta V_{CC}/\Delta V_{IO}$)				70			70		dB
V_{OH}	High-level output voltage	$I_{I+} = 0$, $I_{I-} = 0$	R_L = 2 kΩ, V_{CC} = 30 V, No load	13.5	29.5		13.5	29.5		V
V_{OL}	Low-level output voltage	$I_{I+} = 0$, R_L = 2 kΩ	I_{I-} = 10 μA,		0.09	0.2		0.09	0.2	V
I_{OHS}	Short-circuit output current (output internally high)	$I_{I+} = 0$, $V_O = 0$	$I_{I-} = 0$,	-6	-18		-6	-10		mA
	Pull-down current			0.6	1.0		0.5	1.3		mA
I_{OL}	Low-level output current‡	I_{I-} = 5 μA, V_{OL} = 1 V		5			5			mA
I_{CC}	Supply current (four amplifiers)	No load			6.2	10		6.2	10	mA

† All characteristics are measured under open-loop conditions with zero common-mode voltage unless otherwise specified. Full range for T_A is −40°C to 85°C for LM2900, and 0°C to 70°C for LM3900.
‡ The output current-sink capability can be increased for large-signal conditions by overdriving the inverting input.
NOTE 4: These parameters are measured with the output balanced midway between V_{CC} and ground.

operating characteristics, $V_{CC\pm}$ = ±15 V, T_A = 25°C

PARAMETER		TEST CONDITIONS		MIN	TYP	MAX	UNIT
SR	Slew rate at unity gain	Low-to-high output	V_O = 10 V, C_L = 100 pF, R_L = 2 kΩ		0.5		V/μs
		High-to-low output			20		

TEXAS INSTRUMENTS
POST OFFICE BOX 655012 • DALLAS, TEXAS 75265

A-6. LM2900/3900, quad Norton (Texas Instruments)

LM2900, LM3900
QUADRUPLE OPERATIONAL AMPLIFIERS

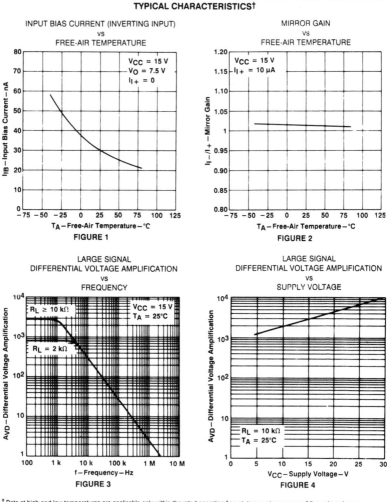

† Data at high and low temperatures are applicable only within the rated operating free-air temperature ranges of the various devices.

POST OFFICE BOX 655012 • DALLAS, TEXAS 75265

A-6. LM2900/3900, quad Norton (Texas Instruments)

LM2900, LM3900
QUADRUPLE OPERATIONAL AMPLIFIERS

TYPICAL CHARACTERISTICS†

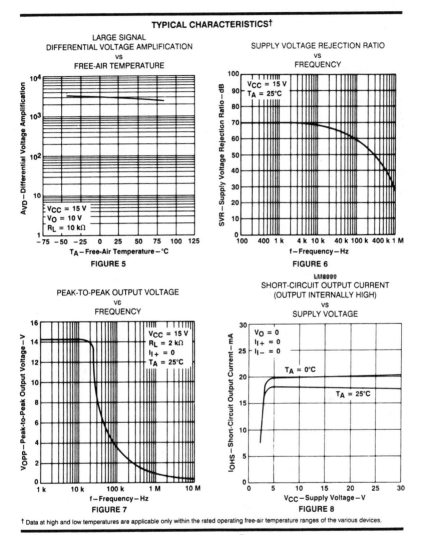

† Data at high and low temperatures are applicable only within the rated operating free-air temperature ranges of the various devices.

POST OFFICE BOX 655012 • DALLAS, TEXAS 75265

A-6. LM2900/3900, quad Norton (Texas Instruments)

LM2900, LM3900
QUADRUPLE OPERATIONAL AMPLIFIERS

TYPICAL CHARACTERISTICS†

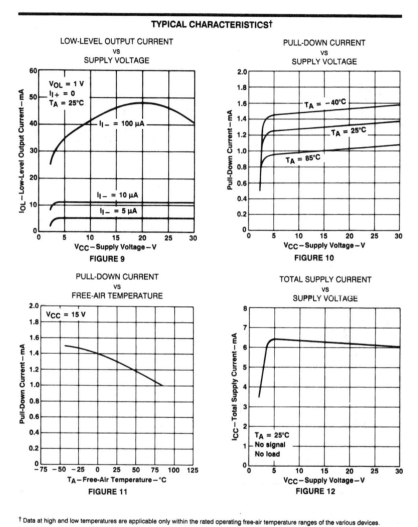

† Data at high and low temperatures are applicable only within the rated operating free-air temperature ranges of the various devices.

POST OFFICE BOX 655012 • DALLAS, TEXAS 75265

A-6. LM2900/3900, quad Norton *(Texas Instruments)*

LM2900, LM3900
QUADRUPLE OPERATIONAL AMPLIFIERS

TYPICAL APPLICATION DATA

Norton (or current-differencing) amplifiers can be used in most standard general-purpose op-amp applications. Performance as a dc amplifier in a single-power-supply mode is not as precise as a standard integrated-circuit operational amplifier operating from dual supplies. Operation of the amplifier can be best be understood by noting that input currents are differenced at the inverting input terminal and this current then flows through the external feedback resistor to produce the output voltage. Common-mode current biasing is generally useful to allow operating with signal levels near (or even below) ground.

Internal transistors clamp negative input voltages at approximately −0.3 V but the magnitude of current flow has to be limited by the external input network. For operation at high temperature, this limit should be approximately −100 μA.

Noise immunity of a Norton amplifier is less than that of standard bipolar amplifiers. Circuit layout is more critical since coupling from the output to the noninverting input can cause oscillations. Care must also be exercised when driving either input from a low-impedance source. A limiting resistor should be placed in series with the input lead to limit the peak input current. Current up to 20 mA will not damage the device but the current mirror on the noninverting input will saturate and cause a loss of mirror gain at higher current levels, especially at high operating temperatures.

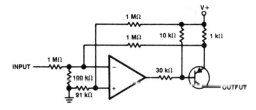

$I_O \approx 1$ mA per input volt

FIGURE 13. VOLTAGE-CONTROLLED CURRENT SOURCE

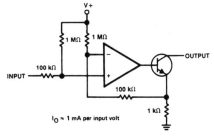

$I_O \approx 1$ mA per input volt

FIGURE 14. VOLTAGE-CONTROLLED CURRENT SINK

A-6. *LM2900/3900, quad Norton* (Texas Instruments)

Appendix B

Standard Resistor and Capacitor Values

RESISTORS

The following ±5 percent standard decade values are available. Those marked with * are the ones most readily available from electronic suppliers.

1.0*	1.8*	3.3*	5.6*
1.1	2.0	3.6	6.2
1.2*	2.2*	3.9*	6.8*
1.3	2.4	4.3	7.5
1.5*	2.7*	4.7*	8.2*
1.6	3.0	5.1	9.1

To obtain standard resistance values, multiply preferred number from decade table by powers of 10. Standard values are available from 10 Ω to 22 MΩ.

The following ±1 percent values are available, but at a higher cost.

10.0	12.1	14.7	17.8	21.5	26.1	31.6	38.3	46.4	56.2	68.1	82.5
10.2	12.4	15.0	18.2	22.1	26.7	32.4	39.2	47.5	57.6	69.8	84.5
10.5	12.7	15.4	18.7	22.6	27.4	33.2	40.2	48.7	59.0	71.5	86.6
10.7	13.0	15.8	19.1	23.2	28.0	34.0	41.2	49.9	60.4	73.2	88.7
11.0	13.3	16.2	19.6	23.7	28.7	34.8	42.2	51.1	61.9	75.0	90.0
11.3	13.7	16.5	20.0	24.3	29.4	35.7	43.2	52.3	63.4	76.8	93.1
11.5	14.0	16.9	20.5	24.9	30.1	36.5	44.2	53.6	64.9	78.7	95.3
11.8	14.3	17.4	21.0	25.5	30.9	37.4	45.3	54.9	66.5	80.6	97.6

These standard values are available from 10 Ω to 22.1 MΩ.

CAPACITORS

In general, capacitor values follow the standard decade values for ±10 percent resistors. The values listed below are in microfarads, and those marked with * are the ones most readily available from electronic suppliers.

.001*	.0033*	.01*	.033*	.1*	.33*
.0012	.0039	.012	.039	.12	.39
.0015	.0047*	.015	.047*	.15	.47*
.0018	.005	.018	.05	.18	.5
.002	.0056	.02	.056	.2	.56
.0022*	.0068*	.022*	.068*	.22*	.68*
.0025	.0075	.025	.075	.25	.75
.0027	.0082	.027	.082	.27	.82

Appendix C

Required Parts and Equipment for the Experiments

Resistors—1/4 Watt Minimum

Quantity	Value	Quantity	Value
2	100 Ω	1	30 kΩ
3	1 kΩ	1	33 kΩ
1	2.2 kΩ	1	39 kΩ
2	2.7 kΩ	2	47 kΩ
4	3.3 kΩ	2	68 kΩ
1	3.9 kΩ	1	82 kΩ
1	4.7 kΩ	2	100 kΩ
1	5.6 kΩ	1	150 kΩ
1	6.8 kΩ	1	180 kΩ
3	10 kΩ	1	200 kΩ
2	12 kΩ	1	270 kΩ
1	15 kΩ	1	330 kΩ
6	20 kΩ	1	390 kΩ
2	22 kΩ	1	470 kΩ
1	27 kΩ	1	1 MΩ

Potentiometers

1	5 kΩ	
1	10 kΩ	(10-turn trimpot preferred)
3	50 kΩ	

Capacitors	
Quantity	*Value*
2	0.0022 µF
2	0.0033 µF
1	0.0047 µF
2	0.01 µF
1	0.047 µF
3	0.1 µF
1	1 µF electrolytic
1	4.7 µF electrolytic
2	10 µF electrolytic
Solid State Devices	
2	1N914 diode
4	1N4001 diode, 50 PIV
1	MPF102 N-channel FET
1	LM318 op-amp (8-pin DIP)
1	LM3900 quad Norton amplifier
1	TL081 JFET-input op-amp or equivalent (8-pin DIP)
2	555 timer (8-pin DIP)
2	741 op-amp (8-pin DIP)
1	7493 4-bit binary counter
Miscellaneous	
1	Breadboarding socket
1	Digital multimeter
1	Dual trace oscilloscope
1	Signal generator
2	0–15 V DC power supplies
1	5-V TTL power supply

Index

A

active filters
  advantages, 145
  circuits, 147-163
  disadvantages, 146
  filter responses, 146
adder amplifier, 9-13
amplifiers
  adder, 9-13
  antilogarithmic, 116
  buffer, 9
  CMOS, 19-20
  current differencing (Norton), 83-93
  difference, 13-16, 78-81
  instrumentation, 16-19
  inverting, 75-77
  inverting summing, 78
  JFET, 19
  logarithmic, 114-115
  noninverting, 39, 73-75
  noninverting summing, 12
  Norton, 83-93
    difference, 91-93
    inverting, 86-88
    noninverting, 88-89
    summing, 90-91
  sample-and-hold, 113-114
  summing, 9-13
analog subtractor, 15
antilogarithmic amplifier, 116
astable multivibrator, 137

B

biasing, current mirror, 84-85
buffer amplifier, 9
bypass capacitor, 71

C

capacitor values, 278
capacitors, 56
  bypass, 71
  compensating, 53
  decoupling, 71

channel separation, 33-34
chatter, 103
circuit input impedance, 36-37
circuit output impedance, 34-35
circuit performance, effects of frequency, 49-53
circuits
   active filter, 147-163
   comparator, 100-113
   nonlinear processing, 95-117
   peak detector, 97-99
closed-loop voltage gain, 5
codes, manufacturer prefix, 20-21
commercial device, 20
common-mode
   bias current, 111
   gain, 46
   input, 46
   input voltage, 4
   output, 46
   rejection (CMR), 33
   rejection ratio (CMRR), 33, 45-48
   voltage, 30
common terminal, 4
comparator
   circuits, 100-113
   double-ended, 108-110
   effects of input noise, 103-104
   limitations and precautions, 112-113
   Norton amplifier, 110-112
   upper/lower switching levels, 106-108
   window, 108-110

compensating capacitor, 53
compensation, frequency, 53-54
Complimentary Metal Oxide Semiconductor (CMOS) amplifier, 19-20
constant-current
   limiting, 125-126
   source, 129
converters, transdiode logarithmic, 115
crosstalk, 33-34
current
   common-mode bias, 111
   differencing (Norton) amplifier, 83-93
   input bias, 31, 40-41
   input offset, 31, 41-43
   mirror biasing, 84-85
   output short-circuit, 32
   regulators, 119-130

D

data sheets, 239-275
   operational amplifier, 26-34
decoupling capacitor, 71
detectors
   level, 101
   polarity, 101
   zero crossing, 101
device identification, 20-22
   date of manufacture, 22
   grades, 20-21
   manufacturer prefix codes, 20-21
   package material suffixes, 22
devices
   commercial, 20

hobby-grade, 20
mil-spec, 20
difference amplifier, 13-16,
  78-81
differential
  gain, 15
  input voltage, 30
differentiators
  basic circuit, 56-58
  effect on wave-forms, 66-67
  explained, 56
  frequency compensated,
    58-62
  gain, 57
double-ended comparator,
  108-110
dual-in-line package (DIP),
  22-23
dynamic parameters, 32-34

E

electrical characteristics, 31
equal-component value VCVS
  filter, 148
experiments, 165-234
  2nd-order low/high-pass
    Butterworth active filters,
    209-217
  active bandpass and notch
    filters, 217-227
  breadboarding, 165-166
  difference amplifier/
    common-mode rejection,
    174-178
  differentiator/integrator,
    183-187
  equipment required, 279-280

format, 167
measurement of operationa
  amplifier parameters,
  178-183
Norton amplifier, 190-195
operational amplifier comparators, 201-207
parts required, 279-280
peak detector, 198-200
performing, 165-167
phase-shift oscillator,
  207-209
precision half-and full-wave
  rectifiers, 195-198
setting up, 166-167
single supply biased inverting
  AC amplifier, 187-190
state-variable filter, 227-233
summing amplifier/averager,
  172-173
voltage follower/
  noninverting/inverting
  amplifiers 168-172

F

feedback resistor, 6
filters
  higher-order, 149-154
  low-pass, 151
  multiple-feedback, 154-158
  Sallen-Key, 147-149
  state-variable, 160-163
  twin-T, 158-160
  universal, 160-163
fold-back current limiting,
  126-127
format, experiments, 167

free-running multivibrator, 137
frequency compensated
   differentiator, 58-62
   integrator, 63-66
frequency
   compensation, 53-54
   effects on circuit performance, 49-53
full-power bandwidth, 52
full-wave rectifier, 97
function generator, 131

G

gain-bandwidth product (GBP), 49-51
gain
   common-mode, 46
   differentiator, 57-58
   integrator, 63
   open-loop, 32, 49-51
gain-setting resistors, 148
generators
   function, 131
   sawtooth, 144
   square-wave, 137-140
   triangle-wave, 140-144
glossary, 235-237

H

half-wave rectifier, 96-97
higher-order filters, 149-154
hobby-grade device, 20
hysteresis voltage, 103-105

I

impedances
   input, 34-37
   output, 34-37
input
   bias current, 31, 40-41
   common-mode, 46
   impedances, 34-37
   offset
      current, 31, 41-43
      voltage, 31, 38-39
   resistance, 32
   resistor, 6
   voltage, 30
instrumentation amplifier, 16-19
integrator gain, 63
integrators
   basic circuit, 62-63
   effects on waveforms, 66-67
   frequency compensated, 63-66
internal
   equivalent circuit, 30
   power dissipation, 30
internally compensated operational amplifiers, 53
inverting
   amplifier, 5-7, 75-77
   comparator, 101-103
   input, 2
   summing amplifier, 78

J

Joint Electron Device Engineering Council (JEDEC), 23
junction field effect transistor (JFET) amplifier, 19

L

large-signal response, 52
level detector, 100
linear amplifier, 2
logarithmic amplifier, 114-115
loop gain, 5
low-pass filters, 151
low supply voltages, biasing, 85-86
lower threshold voltage, 105

M

mil-spec device, 20
monolithic integrated circuit, 1-2
multiple-feedback filters, 154-158

N

negative feedback, 5
networks
   multiple-feedback, 154-158
   parallel-T, 158-160
noninverting
   amplifier, 7-9, 39, 73-75
   comparator, 100-103
   input, 2, 38
   summing amplifier, 12
nonlinear processing circuits, 95-117
Norton amplifier, 83-93
   biasing, 84-86
   biasing with low supply voltages, 85-86
   comparators, 110-112
   difference, 91-93
   inverting, 86-88
   noninverting, 88-89
   summing, 90-91
Norton voltage follower, 89
null adjustments, output offset voltage, 43-45
nulling multiturn potentiometer, 43

O

op-amp, 1-2
open-loop
   gain, 4, 32, 49-51
   mode, 4
operational amplifiers
   active filters, 145-163
   characteristics, 2-3
   data sheet, 26-34
   device identification, 20-22
   differentiators, 55-68
   explained, 1-2
   fundamentals, 2-4
   general description, 30
   integrators, 55-68
   internal equivalent circuit, 30
   internally compensated, 53
   inverting, 5-7
   maximum ratings, 30-31
   negative feedback, 5
   noninverting, 7-9
   operating temperature, 31
   output short-circuit duration, 31
   package styles, 22-23
   precautions, 23-24
   performance considerations, 25-53

power supplies, 4-5
single supply operation,
  69-81
voltage follower, 9
waveform generation,
  131-144
oscillators
  parallel-T, 133-134
  RC, 132-137
  twin-T, 133-134
  Wien-bridge, 135-137
output
  common-mode, 46
  impedances, 34-37
  offset voltage
    contributions to, 37-43
    null adjustments, 43-45
  resistance, 32
  short-circuit
    current, 32
    duration, 31
  voltage swing, 32

P

P-N junction, 114-115
parallel-T
  network, 158-160
  oscillator, 133-134
parameters
  dynamic, 32-34
  static input, 31-32
  static output, 32
peak detector circuit, 97-99
phase-shift oscillator, 132-133
polarity detector, 101
positive feedback, 131

power supplies, 4-5
precision rectifiers, 96-97

R

RC oscillators, 132-137
rectifiers
  full-wave, 97
  half-wave, 96-97
regulators
  current, 119-130
  series, 120-127
  shunt, 127-129
  voltage, 119-130
resistance
  input, 32
  output, 32
resistor values, 277
resistors
  feedback, 6-7
  gain-setting, 148
  input, 6-7
rise time, 34, 51

S

Sallen-Key filter, 147-149
sample-and-hold amplifier,
  113-114
saturation voltage, 32
sawtooth generator, 144
Schmitt trigger, 103-107
series regulators, 120-127
shunt regulator, 127-129
signal mixer, 9
sine-wave signal, 131
single supply biasing, 69-70

slew rate (SR), 32
    limiting, 52-53
small-signal unity-gain bandwidth, 49
square-wave generators, 137-140
state-variable filter, 160-163
static
    input parameters, 31-32
    output parameters, 32
summing amplifier, 9-13
supply voltage, 30

T

time rate of change, 56-58
transdiode logarithmic converter, 115
transient response, 34
transistor outline (TO), 23
triangle-wave generators, 140-144
twin-T
    filters, 158-160
    oscillator, 133-134

U

unity-gain
    bandwidth, 33
    crossover frequency, 49
    frequency, 49
    transient response rise time, 51
universal filter, 160-163
upper threshold voltage, 103-104

V

values
    capacitor, 278
    resistor, 277
voltage-controlled voltage source (VCVS), 147
voltage
    common-mode, 30
    differential input, 30
    follower, 9, 70-73
    hysteresis, 103-104
    input offset, 31, 38-39
    lower threshold, 105
    low supply, 85-86
    output offset, 37-45
    regulators, 119-130
        basic circuits, 120-125
        constant-current limiting, 125-126
        fold-back current limiting, 126-127
    saturation, 32
    upper threshold, 103-104

W

waveforms
    differentation and integration effects, 66-67
    generation, 131-144
waveshapping, 55